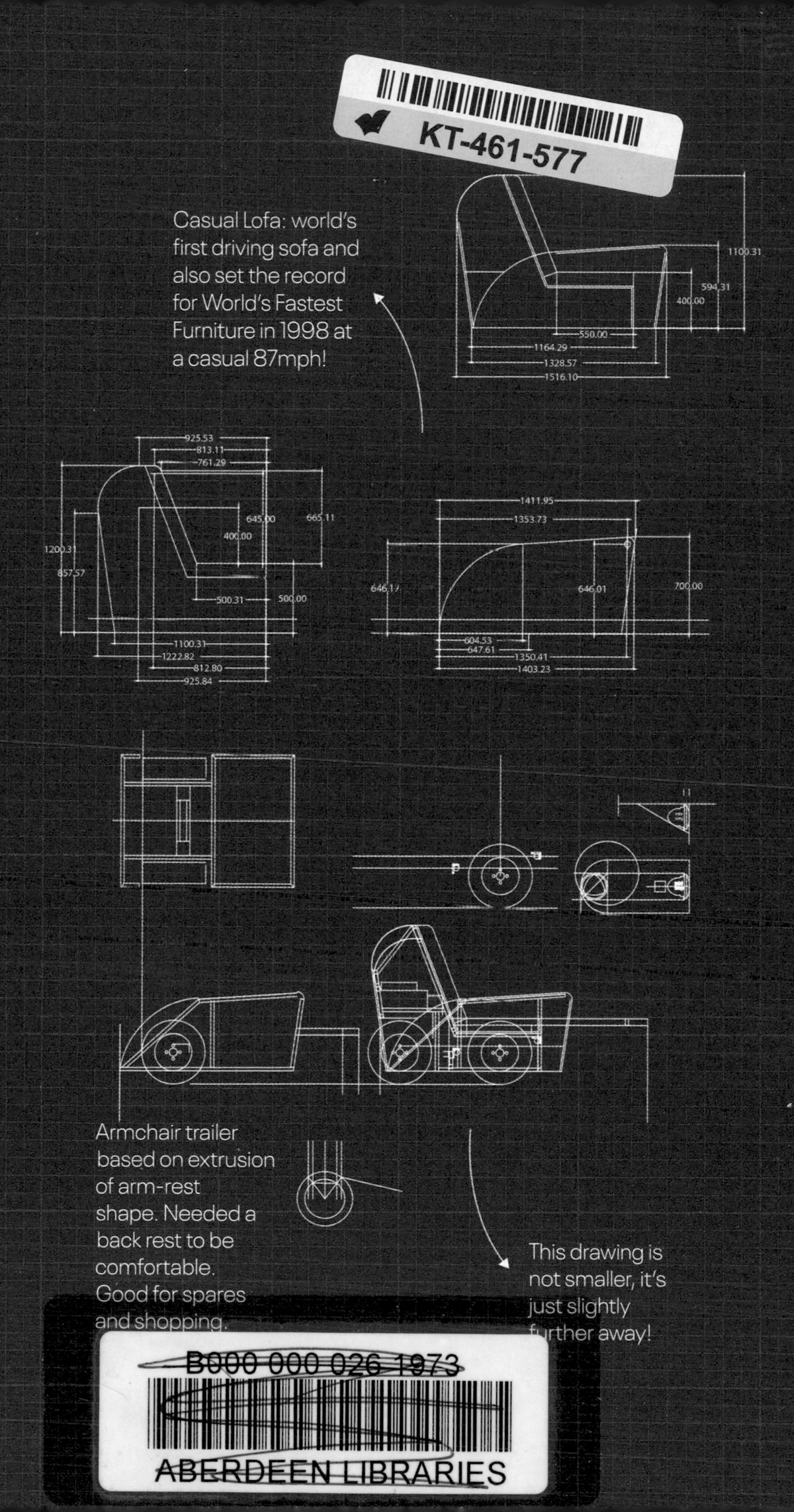

Casual Lofa: world's first driving sofa and also set the record for World's Fastest Furniture in 1998 at a casual 87mph!
Armchair trailer based on extrusion of arm-rest shape. Needed a back rest to be comfortable. Good for spares and shopping.
This drawing is not smaller, it's just slightly further away!

GREASE JUNKIE

GREASE JUNKIE

A BOOK OF MOVING PARTS

EDD CHINA

3 5 7 9 10 8 6 4

Virgin Books, an imprint of Ebury Publishing
20 Vauxhall Bridge Road
London SW1V 2SA

Virgin Books is part of the Penguin Random House group of companies
whose addresses can be found at global.penguinrandomhouse.com

First published by Virgin Books in 2019

www.penguin.co.uk

A CIP catalogue record for this book is available from the British Library

ISBN 978073553541

Printed and bound in Great Britain by Clays Ltd, Elcograf S.p.A.

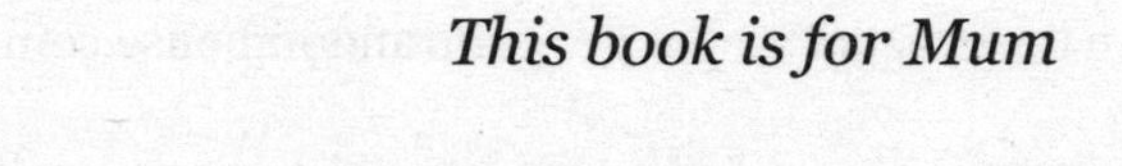

This book is for Mum

CONTENTS

PREFACE

'I AM VERY SORRY TO SAY ...'

A mirage shimmers as the hot Californian sun of June beats down on the winding tarmac of a clear mountain road in Idyllwild. A light breeze caresses the surface of a lake; the faint buzzing of insects and the chatter of birds fill the air. This is a beautiful and rugged part of the world.

'Action!'

The calm is shattered by the deep roar of a V8 engine on song and the mercy cry of tyres losing their grip. Mike Brewer, my *Wheeler Dealers* co-presenter, shifts to a lower gear and crushes the throttle pedal into the thick new carpet of a bright red 1973 Chevy Camaro. The stainless-steel exhaust system sounds fantastic; the unrepentant bark reverberates off the boulders and pine trees that line the road. We tear down a hill and turn into a tight left-hand bend narrowly missing a pair of over-enthusiastic on-lookers pinned against a wall of rock, grit and pine needles blowing into their faces as we pass ...

'And cut!'

The command to pull over and wait in the designated rest area crackles through the radio. Mike turns to me and says, 'This is the last *Wheeler De* ... ' He corrects himself. 'This could be the last *Wheeler Dealers* episode we film together.' I almost don't notice his odd turn of phrase at first but it seems to hang in the air for a moment. Mike has been in an odd mood all day, like a kid who is annoyed that the fart he gleefully produced, stinks. He is clearly frustrated that nobody seems interested or aware that the occasion should be noted or celebrated.

* * *

It was the last day of filming on series 13 and it had been a very difficult year.

On top of everything else, our 1916 Cadillac adventure, the Peking to Paris rally, had just been cancelled by the channel, a week before the car was due to fly out to China. So, rather than going straight to Beijing for the start of an epic adventure, my wife, Imogen, and I stayed on for a couple of extra weeks in our rented house in Newport Beach – a wonderful place to take a well-earned rest. We relished the easy living, did a bit of tidying and then packed our belongings into a storage unit, ready for our return in the autumn. I heard or saw nothing of anyone from the production during this time, or Mike, as he had said he was travelling. So, when I went to catch up and say goodbye to a friend who had worked on the previous series and he said that the production company had been really busy screen testing with some guy from the UK called Ant, I thought nothing of it.

However, as the time for our usual return to filming approached and I still had not heard anything from the production team or the channel, I started thinking something wasn't right; by now we should have been well into pre-production of the next series and I needed to know when I was required back. So, I reached out to the network and requested a meeting.

The next contact Imogen and I had was a conference call with the production company Velocity's boss, Bob Scanlon, the production management and their lawyer.

After the requisite pleasantries and chit chat, they got down to business by saying they were pleased I had requested the meeting as they had wanted to discuss some changes to the production going forward. This was fine by me: in fact, I'd been looking forward to it.

What I wasn't prepared for was the producer's announcement that the show was 'too difficult' to make at the current level, so they wanted to halve the number of episodes filmed per year and, consequently, halve my annual fee. I was (for once) lost for words. Sure, it was difficult, but our previous production company, Attaboy, had managed more episodes with less than half as many staff and, though it was extremely pressurised at times, we had always managed to get a great programme made.

Later, once the reality had sunk in, Imogen and I pointed out that by reducing my pay they had breached the option clause in my contract, thereby rendering the contract null and void. They responded that the smart thing to do would be to just sign a new contract anyway, as they had already been screen testing my potential replacement with Mike and were ready to sign a deal with him if I did not comply. I didn't even hesitate. I had been under an option contract for 13 years, which means I could never have quit even if I wanted to, except for in one, single circumstance: if the company attempted to reduce my fee.

Regardless of the circumstances that led the producers to make the choice they did, and that I might have loved to make more episodes of the show, I had to go for my freedom. I declined Velocity's offer, after which I got an email from Bob Scanlon on 31 August, confirming that there would be a '100 per cent push on Anstead'.

So, rather unexpectedly, I was now free to 'Make My World Bigger', as Discovery's slogan has it. However, the exclusivity obligation in my existing contract meant that I couldn't yet actually act on it or even tell anyone.

At 1pm GMT on 21 March 2017, Velocity/Discovery released the following statement on their social media channels:

> After 13 memorable seasons, Edd China has decided to depart from the *Wheeler Dealers* series in order to pursue other projects. *Wheeler Dealers* will continue for a fourteenth season with Mike Brewer, who will be joined by established automotive expert and master mechanic Ant Anstead.

We waited until we had seen both Mike and Ant's video statements before we recorded mine in our tiny home workshop studio that we had been frantically building and dressing for most of the previous week. We knew the fans would be shocked and wanted to be sure that they understood my underlying reasons for choosing to leave, without going into any of the toxic politics or dropping Mike in the mire.

I felt it was important that I cut through the emotion and addressed a number of obvious issues, and so the wording had to be very specific. I owed it to the audience who had been so supportive and by whose grace the show had become a long running hit. It had to be just right.

The upshot of this process was that it took a number of takes for us to be happy with the statement as a whole – I say 'us' as Imogen plays a huge part in everything I do, and in important times such as this we are both particularly critical of my performance. We ended up with several takes and addendums which all needed to be stitched together in some edit software ... which I suddenly realised we didn't have.

So, I rushed through a load of app reviews online and eventually found some software I thought would be ideal for the job. Twenty quid later, I was busily learning how to feed the video clips from my phone into the app, stick the clips together, ready to finally upload the video onto my YouTube channel. I didn't have time to learn any whizzy editing tricks to make it look even slightly professional, and anyway, we both felt an honest, even slightly raw, hard cut approach was most appropriate. This was direct communication, from me to the fans – no gloss, mood lighting or PR spin required.

Once editing was complete and I had wrestled with all of the correct settings I finally managed to click 'upload' at around 2pm, only an hour after the official statements had gone out. At this point, we felt we had done pretty well; we had created a nice little studio, mastered the rudiments of publishing online video and, at last, we could be open about what we had known for over seven months ...

My statement was relatively brief (by my standards!), to the point, and honest about my feelings about the situation:

> I am very sorry to say, that after 13 years on the show, I am leaving *Wheeler Dealers*. *Wheeler Dealers* is a great car show, reportedly the biggest on the planet at the moment. It is broadcast in about 220 territories around the globe and apparently has several hundred million viewers worldwide.
>
> This success would not have been possible without the huge collaborative effort put in by the production team over the years, but most of all because of the loyalty and appreciation of our fans who have kept watching, in ever increasing numbers around the globe. Because you kept watching it, we kept making it. So, thank you all for that.
>
> Making *Wheeler Dealers* is no easy task, and every episode requires a massive effort from a dedicated team of people. We started, all those years ago, on a niche channel

in the UK with just a small production: Michael Wood and Dan Allum, the founders of Attaboy TV and originators of the show; one production manager; a cameraman; a sound man and Mike and myself. One hundred and thirty five cars later, we have grown to a production team of over 45 staff, and we are now entertaining much of the TV-watching planet.

It's been quite a ride. It was exactly the worldwide popularity of *Wheeler Dealers* and sheer size of the audience, coupled with the lack of product placement and brand endorsements within the show, which meant that Discovery US and Velocity Channel saw an opportunity for further exploitation of the brand and so, after season 12, the show commission was taken over by Velocity Channel in the US, who decided to replace Attaboy TV with Discovery Studios in California.

Unfortunately, on Velocity's first attempt at producing the show, they found *Wheeler Dealers* 'too difficult to make', 'at least in its current format'. In particular, the detailed and in-depth coverage of my fixes in the workshop – what I consider to be the backbone and USP of the programme – are something Velocity feel should be reduced. The workshop jobs are certainly the hardest part of the show to make and reducing their substance and role in the show will save the production considerable time, effort and therefore money. However, this new direction is not something I am comfortable with, as I feel the corners I was being asked to cut compromised the quality of my work and would erode my integrity as well as that of the show, so I have come to the conclusion that my only option is to let Velocity get on with it, without me.

Wheeler Dealers has been an enormous part of my life since 2002, a real roller-coaster of demanding challenges and triumphant moments. It has been a privilege to meet and work with so many great people and, from the workshop to the joy rides, to be allowed to experience so many fabulous cars and amazing locations. Leaving the show at the height of its success has been a really tough decision to make, but I believe the time is right for me to spread my wings. Sometimes you've just got to break free from the confines of your rut and head for the far horizon.

I would like to thank Attaboy TV, Mike, our incredible crew, and everyone who contributed to making such a great programme for all of these years. A special thank you must go to every one of our fans for watching these past 13 series – it has been great having you along for the ride, your appreciation means so much. Mike and Velocity will continue to make *Wheeler Dealers*, apparently with Ant Anstead taking my place in the workshop, so new episodes will be on your screens in due course. Being the new boy is never easy, so please give Ant your support, I wish him the very best of luck. It will certainly be nice to see Mike turn up at the workshop

> with yet another wreck knowing I don't have to do any of the work!
>
> As for me, I am already working on some great new projects which will expand my world in new directions and, as I will have a bit of time on my hands, I'll even start to put some things up on my YouTube channel, so keep an eye out for that. Right now, though, I just feel relieved and exhilarated with freedom and I can't wait to get cracking on the exciting opportunities that lie ahead. For years, Discovery has been encouraging everyone to 'make your world bigger' – and that is exactly what I am now doing.

Our message was ready to send to the world ... and it was at this point that Imogen and I discovered just how slow our broadband was. Uploading this precious four-minute video was apparently going to take several hours! Initially, the predicted upload time kept clicking up rather than down, but there was nothing we could do. We rushed around the house turning off any phone or gadget that might consume valuable data from our paltry allowance, but the tiny packets of ones and zeros still trundled along our phone line in their own sweet time.

Going by the text messages we were getting from Mike and the production team they were becoming increasingly impatient. The upload time clicked down excruciating slowly. All we could do was to take more walks in the garden and make more cups of tea ...

Some three hours later, at 5.20pm GMT, the upload finally completed and the video went live for public consumption.

The YouTube/Edd China channel went into meltdown and we soon had three million views and 20,000 comments to sift through. A new chapter was beginning; now all I had to figure out was what to do next ...

1.

A TEXAS YELLOW BUG

Ever since I was a small child, I have always been fascinated by how everything in the world around me works. Back then, I couldn't resist taking apart everything from the lawnmower to the toaster and my mum's clock radio, just to discover their hidden secrets and what made them do what they do. The most coveted mystery was always my grandparents' cuckoo clock; I was desperate to get my hands on it, but I never quite managed to.

Strangely, my mum didn't quite share my enthusiasm for 'the joy of knowing what's inside', so eventually I had to become quite good at putting things back together too. This led to me being able to fix almost anything.

But before that, I was a Lego Kid. My most treasured possession was the Lego bag – a circular piece of fabric with string that contained all my Lego which would open flat on the floor, giving me access to all of my pieces at once, so I could just get on with building. It never bothered me that everything was in a big jumbled mess, I could always find, with laser-like focus, exactly the piece I needed.

For a special treat, or if I had birthday money, my mum would take me to the local toy shop to buy some more Lego. Every time she would tell me not to open the box until we got home as I was sure to lose pieces in the car, which would ruin the set – and every time I would open the box and pieces would be lost. That didn't ruin the set for me though; I would just build something else, similar or completely different to what the instructions said.

Actually, I never read the instructions. I just looked at the picture and worked it out for myself. I still consider instructions to be helpful but optional, as to understand how something works I need to understand how it goes together, and so I have just sort of developed a sixth sense about it.

I'm also convinced that my heightened spatial awareness and ability (some say preference) to excel in chaos is due to having spent most of my childhood sitting on the floor surrounded by Lego bricks, working on a masterpiece while scanning for the next essential piece. To this day, chaos and uncertainty gives me a deep feeling of well-being and a sense of potential and new possibilities.

The other thing I really loved doing was talking, or chattering as my mother rather uncharitably called it at the time.

She may have had a point. Even before I had any words, I would hold forth in a constant stream of '*adju adju adju adju*' from the moment I woke until I finally went to sleep and my mother could rest her ears. If she thought that was bad though, it did little to prepare her for the furious stream of questions, observations and postulations that followed once I did finally learn to speak.

I was a keen observer of everything around me and would provide an unrequested running commentary on it all, asking ceaselessly, about all the hows, whys and wherefores. I was just wired that way. I guess I still am.

The pinnacle of joy for me was when I could sit on the floor in my grandparents' bay-window, surrounded by my Lego, furiously building while chattering in a constant stream of consciousness about everything I had seen and everything I had learned or pondered recently.

So maybe that's where it all started – furiously building while furiously chattering, all while constantly scanning for the next piece and solving the next problem ...

* * *

Of course, I have become best known for my fascination with motor vehicles of all kinds, and for enthusiastically solving automobile problems.

My first car was a 1303 Texas Yellow VW Beetle with a 1600 cc engine and yellow and blue eight-spoked Empi alloy wheels. It was bought pretty much on a whim but turned out to be a brilliant first step on a lifelong adventure. It looked fantastic and had everything you would expect from an old Beetle - including major rust problems and an engine that kept conking out.

I didn't know that it came with those less-than-desirable optional extras when I bought it but they are most definitely what set me on the road to discovering what cars are all about. Necessity, they say, is the mother of invention - and of learning how to fix a carburettor in the middle of the Dartford Tunnel during rush hour. Without doubt, the near endless to-do list of my first Bug helped set the course for my professional journey, fixing not only VWs but all types of vehicle.

My first automotive memory is thundering across Clifton Common in the rear seat of my Uncle David's 1929 4.5 litre Bentley. Apparently, it was also the first time I had travelled at 100mph. It was a wonderful, exhilarating experience.

My uncle owned a number of lovely old cars but my memory of them all is now a little hazy. As I remember it, the vintage Bentley was being restored by an old boy and pretty much the moment he had finished taking it completely to pieces, he died. I believe the car was eventually sold, in boxes, for a fraction of its true value. Interestingly, some years back, I heard on the grapevine that it had re-appeared, in one piece, somewhere in mid-America. Maybe one day I'll get to drive it at 100mph myself ...

I learned to drive in my mum's Nissan Bluebird estate. Before I was even old enough, I was desperate to get my driving licence. I just couldn't wait to learn. For the two birthdays leading up to my seventeenth birthday I had asked for nothing but driving lessons and, in the end, I received enough money for around 20, so I even got a bit spare for petrol.

When the day finally came, the driving instructor turned up in a ridiculously tiny Mazda 323 Javelin and I could barely fold my six-foot-seven frame into it. After the first lesson, it became apparent that I would not be able to carry on in that miniscule car so I asked if we could maybe continue lessons in my mum's full-sized car, and he was only too happy to agree. He hadn't enjoyed having my elbow lodged in his ribs any more than I had enjoyed bashing into every surface in his car.

So, I ended up taking the lessons in the Bluebird, which meant I was completely at home in the car by the time it came to taking my test. My final lesson was a 'mock driving exam'. To test my emergency stop, the driving instructor explained that I should drive along as normal and the moment he raised his clipboard, I should come to an immediate stop. I set off down one of the many back roads of Farnborough; up went the clip-board and I came to a casual halt.

'That wasn't much of an emergency stop!'

'It wasn't much of an emergency!'

Duly chastised, the split-second the clipboard next went up, I planted my size 14 foot on the brake to such startling effect that the instructor smacked his head on the windscreen. I was ready.

Once I passed the test, there was no stopping me; I could go anywhere and do anything – as long as I was allowed to borrow the car. Over the next year, as my mates and I started planning all the excellent adventures we would have as soon as our A-level exams were over, it became apparent that I needed to get my own wheels.

We had a deadline. Summer was now only a few weeks off and we wanted to get down to the beach, so I enlisted the help of my mates to find the right motor. All I knew was that I wanted something a bit different, something unique that would stand out from the crowd and would look cool with our windsurfing boards on, but, beyond that, I hadn't thought it through very much at all. I really knew nothing about how cars worked, I just needed a car.

The lads and I trawled through the classified ads in car mags, local newspapers and *Autotrader*, or *Thames Valley Trader* as it was called back then. We used every spare moment, even when we really should have been revising between exams. We compiled a list of contenders, but then a Jeep with big, chunky tyres and shiny wheels caught my eye. It seemed to capture the spirit of our imminent freedom. We phoned the number in the advert, sneaking out during lessons to use a friend's landline phone, but we could never get through to the guy who was selling it. Eventually, after a week and a half, we finally spoke to him only to find that he had sold the car over a week before. Gutted!

In the same paper, however, there was this 1303 'Cal look' Beetle in bright 'Texas Yellow', with yellow and blue eight-spoked Empi alloy wheels for sale. I had no idea what 'Cal look' was, or that the 1303 is a later model Beetle with details such as a curved, rather than flat windscreen and a padded, rather than metal dashboard. Or that it has MacPherson strut suspension and IRS (independent rear suspension), making for better handling. I just thought it looked great in the photo. This was a cool car.

Having arranged a time to view it, I set off with a couple of mates to darkest Uxbridge. Nobody else had their licence, or a car, so my mate Graham's older brother Nigel took us in his parents' trusty Honda Accord. I had a bit of an inheritance from my grandmother and I was ready to blow it all on a car – about £2,000 as I recall.

When we got there, we had no idea what we were looking at, and actually did the whole tyre-kicking thing. We'd heard about that. And we'd heard about the shock absorber thing, where you lean on the wing, push the car down and it has to settle back to where it was within one-and-a-half bounces, otherwise it's time for a sharp intake of breath and a bit of head shaking. We did that at each corner. We must have looked like total numpties.

We continued with our unconvincing portrayal of seasoned car traders and inspected the underside, the tyres again, the paperwork ... but the truth was that I wanted the car from the moment I clapped eyes on it. The bodywork had been repainted relatively recently and the engine was clean and nicely detailed. It was, overall, a cool car, and given that we had spent a fair amount of time checking out car prices in the papers, it seemed like a bargain, too.

I decided that I needed it, and with the cash burning a hole in my pocket, I simply had to buy it. No checking the thing out properly, no real knowledge of how the car should look, sound or drive – so, essentially, the worst possible way to buy a car. After a rudimentary test-drive I handed over the £1,500 asking price and we drove the car away.

We only just made it onto the M25, before it ran out of petrol and I discovered the first issue on the car – the fuel gauge wasn't working.

As we were in two cars, two mates went off to fetch a can of petrol and a couple of us were left on the hard shoulder, hanging off the side of the car, 'surfing' on a running board as trucks went past and rocked us. That was when I found out that the running boards on the Beetle – the ledges that sit on the sills below the doors – were aftermarket running boards which really aren't very substantial at all. You can't stand on them like hit men in a 1930s gangster movie, because they're not designed to take that sort of load. Volkswagen's own running boards could just about take your weight, provided that the car wasn't too old or rusty, but these things were really flimsy and just buckled. The first damage I did to the car, therefore, was to break a running board on the way home from buying it.

Eventually we got some petrol in the thing and I got it home. My immediate priority was to rush into Godalming and use up the money I had left to buy tax, insurance and a car stereo. I also joined the AA, which turned out to be a sound investment. So, with only a few exams left to get through, the beach trip was on.

* * *

My mates Steve, Alison and Sue and I strapped windsurfing boards to the roof, loaded up the summer sounds mix tape and headed for the coast. We felt on top of the world and had a fantastic first day out on the beach at Hayling Island until, on the way back, my Bug started spluttering and coughing even more than a Beetle usually does. Finally, the 1600cc engine cut out completely and we coasted to a stop at the side of the road. There was fuel in the tank. Not much, but enough, and having walked around the car a few times, scratching my head, looking in the engine bay and giving the car a few good shoves, it started up again.

It was then that I realised that I knew absolutely nothing about cars and I needed to start getting my head around how they worked, the things that could go wrong and how to fix them. Filling up the fuel tank seemed to make it run better for a while but the same thing kept happening – the Bug would be running fine for a while and then it would conk out. I eventually noticed that it tended to happen on inclines, particularly when the fuel level was low.

I've still got the scar from when the bloody thing broke down on the A3 and I tried to push it up a hill. On the hinges of the rear boot lid, which of course on the Beetle is the engine cover, I had what are called 'deck lid stand-offs' which allow the engine cover to sit out from the bodywork. This looks cool and is supposed to help keep your engine cool too by scooping in extra air. Of course, as the Beetle engine is air-cooled, an influx of cool air is vital, but the merits of deck lid stand-offs are debatable as, obviously, Ferdinand Porsche and his team at VW had already thought about how to prevent their engine overheating.

The way this works is that a Beetle's cylinders are covered in cooling fins, just like on an old air-cooled bike or early plane engine. On the back of the alternator is a centrifugal fan inside a housing which sucks cold air in from outside and then blows it through the fins to remove the engine heat. To maximise the cooling efficiency, the engine is surrounded by a tin jacket which guides the air around and past the cylinders and heads and then out of the bottom of the engine bay.

V-dub engines often suffer from overheating when a part of this tin gets mislaid, allowing the air to just recirculate within the engine bay, getting hotter and hotter, until the metal of the engine expands with the heat and eventually grinds to a halt. A quirk of the flat four's design means that cylinder three, which sits under the oil cooler where it's slightly warmer, is often the first to seize up. If you are lucky though, giving the engine enough time to cool down will free things up and your Bug will run again.

So, I had these totally unnecessary deck lid stand-offs and the top edge of the lid itself had quite sharp edges because it was never designed to sit in that sort of open position. Pushing the car up that hill on the A3, trying to get it to a layby, I slipped and gashed my wrist. I wasn't exactly mortally wounded but there was blood everywhere, which looked especially dramatic smeared all over that bright yellow body work. I tied a rag around the wound and carried on struggling when, thankfully, some guys in a camper van stopped and gave me a tow to the top so I could roll down the other side. Pointing downhill, with all the movement having stirred the tank, the Bug started up, no problem at all.

By now I was pretty sure that the problem lay with the fuel tank, so the next time I sputtered to a stop I disconnected the fuel hose from the mechanical fuel pump and blew the fuel back through the hose into the tank. This cleared the blockage and gave me a few more miles of smooth running. At last, the problem was identified and fixable! So, I promptly put off doing anything about it a little longer, until I finally got fed up of gasoline gargles.

It was time to take out the fuel tank and deal with the problem. It turned out that the previous owner had managed to get silicone sealant into the fuel tank. There was a breather pipe missing and, rather than fit a new one, they had clearly tried to seal the hole using silicone, some of which had slid down into the fuel and never really gone off. It lurked near the fuel outlet, obstructing the flow whenever there wasn't enough petrol to keep it sloshing around, or when an incline coaxed the silicone into the perfect position to seal the fuel's exit.

After fishing out every last scrap of the offending sludge and an excessive amount of flushing out for good measure, I replaced the tank and the missing breather pipe. This was the first major fault I had encountered on my Bug, an intermittent fault which had taken ages to figure out. It was a real baptism of fire but the sheer joy of having triumphed was thrilling. Still, I wasn't exactly in any hurry to go scraping my knuckles on further repair work, which was a shame, as there was an endless stream of jobs to come.

Shortly afterwards, I took the Bug for its MOT test, which it duly failed. The problem was rust in critical places such as near suspension and seatbelt mounts. It was now clear why the car had been such a bargain. Anyone who had given it a proper inspection would have known that it needed a fair amount of remedial welding. While the previous owner had made the car look pretty, it wasn't actually roadworthy. I had no job, no money left and no idea how I was going to get my pride and joy back on the road.

Luckily for me, there was a small garage down the road from us and the chap did a bit of welding. He took pity on me and told me that if I took the car apart and got it ready, he would do the welding on the understanding that I would pay him when I could. That worked for me. Welding was needed on the sills, the heater channels, the floorpan and the battery tray under the rear seat. So, I removed as much as I could to get the best possible access; the doors came off, the seats, the spare wheel, the rubber floor mats, any potential fire hazard ... if it was removeable, it was stripped out.

Stripping the car made me aware of how important it is to use the right tool for the job – in this case, not using imperial tools on a metric car. The only tools I had found in the garage were my dad's imperial tools for working on his Minis. Mum and Dad each had a Mini and they'd taken them both apart to make one good one. I vaguely remember that at least one of those cars left our drive on the back of a lorry so it seemed right to continue the tradition of having a go, come what may.

The imperial Mini tools did not easily fit the metric Beetle nuts and bolts. After several skinned knuckles and much swearing, I had to go out and buy my first tool, a 13mm spanner. Unfortunately, by this time, I had rounded off the corners of the nuts holding the exhaust system onto the engine. My new spanner had freed off one or two but I needed a plan B. Thankfully, I discovered some very clever sockets that work on the surface of the nut rather than the corners and the old heat exchanger was off in no time. There is no getting away from it, you always need the right tool for the job. Each time I came across a new problem I slowly built up my tool kit, one tool at a time.

The garage wasn't very far away, so I thought I might as well drive the car down – even though it was missing a few things, I figured it was probably mostly legal. As I took off down my road, the acceleration rather caught me by surprise. The car was just so much faster without all of that extra weight, and the wind rushing through the gaping holes where the doors should have been added further excitement to the experience. This definitely needed further investigation.

We put the car up on axle stands to get a better look and assess the situation. The MOT tester had marked out the offending areas with yellow chalk but, following a bit of grinding, it became clear that there was even more rot than we had thought, much of it hidden behind bad repairs. We got to work cutting away the rust and made new repair pieces, out came the oxy acetylene torch and away he went. Things seemed to be going well and there were only a few minor fires for me to put out with a wet rag.

This was clearly not the first time my Beetle's metal moth infestation had needed attention. There was a large repair patch at the front of the car, on the narrow part of the floorpan that supported the suspension's lower control arms, so it was quite an important area. It looked like the previous owner had gone to the trouble of making a nice, neat repair with impressively tidy welds until we took a closer look and it turned out those welds were flammable and the patch wasn't steel!

The tin or aluminium had been knocked into shape and then 'stuck' in place with mastic. We prised it off only to find a huge rusty hole. It was amazing that the front of the car hadn't just snapped off. The suspension was moments away from collapse. That was my first lesson in how you can't always trust a car salesman. It was also my first introduction to welding. It was clear I needed to learn that skill.

Once all the welding was done, I put the car back together, took it back to the MOT station and paced around nervously outside, waiting for the verdict. It passed. I raced home with the good news to find Mum getting ready to pick up my sister, Clare, from school. At, first Mum was rather reluctant to go in the Beetle instead of her more suitable Bluebird, but I soon convinced her that my freshly certified car would be fine. After all, it's not as if the wheels would fall off or anything ...

And then, about half a mile into our journey, they did. Or, at least, one of them did. We were on a short stretch of dual carriageway outside Farnborough when there was a strange, juddering, vibrating noise and the nearside front wheel was suddenly bouncing along the road beside us. It is a little unnerving to see one of your car's wheels racing you down the road. I was amazed that the car hadn't totally collapsed and spun off into the Armco!

I gingerly steered the car towards the inside lane of the dual carriageway while trying to match the speed of the escaped wheel. I was desperately trying to herd the wheel to keep it from bouncing across onto the other side of the road and causing an accident. Steering the car with only one front wheel was a challenge but at least Mum's bird-like stature was no match for my weight, which was on the side of the car that still had a full complement of wheels.

Meanwhile, we were literally grinding to a halt, leaving an impressive trail of sparks behind us as the bottom ball joint of my MacPherson strut suspension was grinding along the road and melting the yellow lines. We weren't going to get very far like that and, thankfully, the errant front wheel soon came to rest, with us alongside it.

You'd have thought Mum would have been impressed – my driving skills had clearly saved the day. Strangely, that's not the reaction I got. I retrieved the wheel and even managed to find one of the wheel bolts, so I only had to borrow one from each of the other wheels to get the missing wheel back on its hub and get us home. Mercilessly, Mum did not let me live the incident down as we set off in her car to pick up my sister.

Now that I had got my car running properly and roadworthy, I decided to have a go at my first bit of customising.

The 'Cal' or 'California look' is more of a loose collection of ideas rather than a strict set of unwritten rules. True to its low-rider 'Cal-style' origins, the car's ride height should be lowered to the ground – 'slammed' or 'dropped into the weeds' – by adjusting or modifying the suspension. The paint should normally be a non-metallic pastel colour and most, if not all, of the trim and chrome should be removed, sometimes including the bumpers, to give it a stripped back, 'shaved' look.

The previous owner hadn't lowered my car, but he had made a start by removing some of the trim and adding hooded headlights, or eyelashes as we called them. To go full-on Cal look you could even remove the door handles and use solenoids inside the door to operate the locks, like the ones I fixed on the 1952 DeSoto on *Wheeler Dealers.*

This seemed like a good place to start; so, inspired by an article in *VolksWorld* or *Custom Car* magazine, I found myself a second-hand VW starter solenoid, some spare throttle cable and a small pulley and I was ready to go smooth. It was the first custom work I had ever done, and as a result I picked up a top tip that would come in handy many years later on the show, when I did a bit of detail work on a Ford F1 rat rod truck.

When removing the door handles, you are left with a couple of holes and an indentation in the bodywork where your knuckles go. For a true Cal look, I needed to remove any sign of them ever being there, so I cut and slightly rolled a piece of steel to cover the dimple perfectly, matching the contour of the door skin. Having ground off the paint on the door where the parts would be welded, I held the new steel in place with a magnet before slowly spot-welding the pieces into place, making sure not to let the metal get too hot as that can easily distort the door skin when the weld cools and the metal contracts. The final stage before filler and paint was to grind off all of the excess weld leaving a smooth blended surface. Job done!

I was really chuffed with the end result. My first bit of car customising had worked a treat and looked great. Now, I am not so good at getting around to regularly washing my cars, so I was a bit peeved when, some days later, following a heavy rain storm, I noticed that my car's windows and windscreen were looking rather dirty and brown; you might even say rusty. It turned out that the molten metal sparks from the grinding and weld spatter had fused with the glass and then oxidised, as steel left out in the elements is wont to do. Not brilliant at the time, but it did at least teach me to always cover glass when welding or grinding near it, and I turned my mistake on the Bug into a great effect all those years later when I 'Eddtched' the *WD* logo onto the F1 truck window using the same technique.

As cool as my Bug was now becoming, it was not the ideal car in which to go on a family holiday to the South of France. So, Mum, Clare and I set off in my mum's Nissan Bluebird estate and I did the lion's share of the driving. It was the farthest that I had ever driven and somewhere along the way I managed to figure out how to change gear without using the clutch. Get your engine speed and road speed about right and you can knock the gear lever out of one gear and into the next. Great when you get it right. A little crunchy and not very good for your gearbox if you don't.

We were staying near Biarritz, in a place called Cambo-les-Bains with a group of family friends, and we had a fantastic time visiting the beaches in the area and checking out the towns. I was fascinated by the way nobody put their handbrake on when parking, so when somebody else came along to park, they could just shunt all the other cars along to make space. One day, we came across a Citroen with all four of its wheels off the ground, suspended on the bumpers of the cars in front and behind it. By one of the beaches, I also saw a gorgeous, bright lime green Karmann convertible Beetle towing a matching speedboat and that gave me an itch I only got round to scratching many, many years later on the show, when we gave a particularly lavish paint job to a 1962 Cadilllac Coupe de Ville.

One day, as we were parking by the beach, a bunch of young lads drove past and squirted us with water pistols. Great gag, we thought, and at the first opportunity we armed ourselves with some water pistols. They were very cheap and plasticky but I suppose they did look a little like real guns. For much of the rest of the summer, if any of our friends found themselves unarmed, they got squirted.

The joke didn't tire for weeks. Well, not until we got arrested for attempted armed robbery.

It was a fabulous blue-sky summer Saturday – scorching hot, the sort of day that makes you wish you had bought a convertible. It was a day for lazing around doing nothing in particular, but I had been working since early morning, helping out with a loft conversion. Doing physical labour in a loft space with the sun toasting the roof tiles is like working in a sauna, so I had spent most of the day in just boxer shorts and trainers. When my mates started to assemble to figure out what we were going to do that evening there was the usual aimless discussion that occasionally broke out into water pistol battles with the arsenal of squirters. Then my friend Nick remembered that he had to get home to feed the cat. All five of us piled into the Bug and set off.

As usual, I had about a teaspoonful of petrol in the tank, and so we had to pay a visit to a filling station. Hilariously, en route, someone had a go at squirting a group of pedestrians on their way to the pub. We were far more cautious going through an army checkpoint – Aldershot was the home of the British Army, and was a constant target for the IRA, so we were used to seeing soldiers out on the streets. There was some sort of IRA anniversary going on, so the squaddies were on high alert and armed with real guns. We kept ours out of sight on the floor as we crawled through the various stages of their traffic control, chatting and having a laugh with them as they wafted mirrors around under our car checking for bombs.

At the filling station, there was another heated discussion, this time about who should get out to fill the car up, who should go into the shop to pay and who was going to stump up the cash. I certainly wasn't getting out of the car as I was still only dressed in boxer shorts, and if I was going to be ferrying them around that night, I didn't see why I should pay for the petrol.

Between them, the boys managed to scrape together £5 in loose change. Logic prevailed and the front-seat passenger, Graham, was chosen to do the filling up. Now free to get out from the rear bench seat, Aaron decided he might like a snack, so stupidly, and rather typically, he judged that the smart thing to do was take a water pistol with him to avoid getting squirted on his return.

Graham finished refuelling and went in to pay. A fan of Schwarzenegger and his work, Aaron entered the shop with the gun held in both hands up by his head. Distracted by the confectionery display, the boys lamented the change of name from Marathon to Snickers, and how much more fattening the new Kingsize Mars bar might be than the standard size.

Slightly perturbed, the girl on the till asked them whether the gun was real. 'Of course not!' was the slightly perplexed reply and, to allay her fears, Aaron did the decent thing and gave Graham a little squirt as they explained that it was just a water pistol. 'If it had been real, I'd have had to put you on camera,' the cashier warned them. 'But, OK, no worries.'

The boys paid for the petrol and the chocolate, piled back into the car and we drove off to enjoy the rest of our Saturday night.

On Tuesday of the following week, I drove into Aldershot to visit a car stereo shop because my radio wasn't working. I needed a new aerial. I pulled up outside the shop in my bright yellow Bug and was sitting on yellow lines trying to decide whether to risk a quick dash into the shop before a traffic warden turned up, or whether I should invest a few coins in the parking machine at the car park just up the street and walk back to the shop.

A police car drew up behind me. That made my decision for me: I needed to move on. An instant later, another car pulled sharply in front of me and reversed back right to my bumper. Really? What was he playing at? Hadn't he seen the police car behind me? He was going to get me a parking fine! The next thing I knew was a very loud voice screaming at me: 'Armed police!

Don't move! Stay still! Armed police!' A policeman was suddenly pointing a very large gun straight at my head.

Suddenly, there were policemen everywhere and I was hearing all sorts of commands. 'Stay still!' 'Hands on the steering wheel!' 'Freeze!' 'Switch off the engine!' 'Don't move!' 'Get out of the car!'. I was trying to work out how to get out of the car without moving when a policeman reached in and grabbed the keys out of the ignition. A hand on my shoulder yanked me out of the car.

It seemed, to say the least, quite a strong reaction to a potential parking misdemeanour. I was totally shocked at first, a bit numb, then I was really scared, and then I started thinking: 'Hang on a minute. This is all a bit silly. There's been some kind of mix-up. They've got the wrong bloke.' At that point, however, it seemed the best policy to keep my mouth shut.

I was manhandled into the back of a police car closely followed by one of the officers, who told me: 'I am arresting you under the charge of attempted armed robbery.' Before I had a chance to process what was going on, I was driven around the corner to Aldershot police station and another officer followed us in my Beetle. After some form-filling and waiting around I was bundled back into a police car and transferred to Farnham CID for questioning.

While this was all going on, it became clear that the police didn't think they had got the wrong bloke. They had arrested precisely the criminal mastermind that they had been looking for since Saturday – the driver of the vehicle used by a gang of armed robbers. Apparently, the girl in the petrol station had mentioned the 'incident' to her boss, who had complained to the police, leaving off a few crucial details and maybe adding a few for good measure. I was put in a room with a senior officer, DC Boon, who, I think, was starting to suspect that this had all been a major balls-up.

Our conversation went a little like this:

'You understand why you have been detained?'

'No, not really.'

'We have reason to believe that you were involved in the attempted armed robbery that took place on Saturday.'

'What armed robbery? I don't know anything about an armed robbery.'

'It was on the front page of the local paper.'

'I haven't seen the local paper.'

'It was on the radio news.'

'My aerial's broken, that's why I was at the stereo shop.'

'You seem to have an answer for everything.'

'You generally do when you're telling the truth!'

Once I realised what was going on, I began to find it all quite amusing. Uncomfortable, not laugh-out-loud funny, but mildly amusing. I explained what had really happened and reasoned with my interrogator that, if we had actually staged an armed robbery, surely I wouldn't have used my own car? Surely we would have filled up the tank, rather than just taking a thimble-ful? And surely we would have emptied the till rather than picking out a measly couple of bars of chocolate? Plus, surely no self-respecting armed robber would stump up his own cash and actually pay for the fuel and snacks? That's not really a robbery at all, is it? At worst, it's an armed purchase!

I had to give the names and addresses of everyone who had been in the car with me and I was then locked in a cell in the custody suite while they checked out my story. They took my sunglasses, my shoelaces and my belt. When I asked why they were taking my belt they said it was because I might use it to hang myself. I thought, 'Why would I try to hang myself with my belt? Have you seen how long my jeans are?' I stopped myself from actually saying that in case they decided it was a reasonable argument and made me hand over my jeans as well.

The police set about calling my mates, who all thought it was some kind of wind-up. Even their parents weren't convinced it was really the police. I just snoozed in the cell while we waited for everyone to be interviewed and eventually it became clear that we were more innocent victims than ruthless villains, because the stories that the police had gathered from our accusers and various witnesses simply didn't add up.

According to witness statements, we were supposed to have soaked a group of pedestrians, gone through the checkpoint at great speed, all been under six feet tall with 'chemically lightened hair' but, most outrageously, witnesses had accused my Bug of having *cream coloured* running boards!

Now, hang on; my Beetle had *black* running boards!

Sure, the other things were preposterous. Soaking anyone with a squirt gun is virtually impossible, especially if you're passing them at 30mph. Speeding through that multi-stage checkpoint would simply not have been possible and if we *had* attempted to blast through, our evening would have ended in a hail of bullets. Now that *would* have made a front page story in the local paper.

Amusingly, we were all comfortably over six foot and not one of us was blonde – although one *was* ginger.

Still, how dare they get the description of my Bug wrong!

They did, however, get the registration number right. Tracking down the car from the licence plates resulted in a dawn raid on the home of the bloke in Uxbridge that we had bought it from. That must have been the alarm call from hell. It transpired that I had neglected to send off the V5 form to the DVLA to advise them of the car's change of ownership. Well, served him right for selling me a dodgy motor.

Footage from the security cameras at the service station eventually backed up our story, and, after all the evidence had been considered, the charges were dropped. However, we were told very sternly that as a result of this incident we now had some kind of unofficial police record and that we were 'bound to keep the peace' for a year if we wanted it to stay off our records (although I think only a magistrate can actually do that).

The police were clearly trying to save face, but we were definitely on our best behaviour for a while after that, and I was to have many more miles of fun in my Texas Yellow Bug – until the day a Rover pulled out in front of me on the A13.

I was driving through London, from Rainham to Farnborough. I was somewhere near Limehouse, where the road is a dual carriageway. Traffic was uncommonly quiet.

I noticed a blue Rover 200 waiting to join from a side road on my left. I moved into the outside lane to give him plenty of space. Suddenly, at the last moment, the Rover lurched out in front of me, straight across both lanes. The old guy who was driving it must have been riding the clutch and his foot slipped off, causing his car to shoot forward at precisely the wrong moment.

I had nowhere to go. The front of my Bug piled into the side of his car.

It is amazing, the things that flash through your mind when you know that you're going to crash. I think that your brain speeds up, like a switch has been flicked that allows you to think faster and maybe save yourself. Certainly, everything else seems to be going more slowly. Your senses are heightened, you can see really clearly and the noise of the impact always seems really loud, even if it is only a little bump. This time, though, it *wasn't* a little bump.

Beetles were not really designed for 'full-size' people like myself, so I was forced to adopt a kind of praying mantis position when driving, my knees wedged up against the padded metal dashboard, either side of the steering wheel.

As the Bug rapidly decelerated during the impact, my own inertia forced my knees into the dashboard, the bending metal helping to bring me to a stop inside the car. My knees left dents in the dashboard a couple of inches deep.

My first thought was for the old guy and his wife. I had to make sure that they were OK. I got out of the car to rush over and check on them, and immediately crumpled to the ground. My knees weren't injured, a bit bruised maybe. Though heaven knows what might have happened to my legs if they had smashed into the dash from even a short distance. I was all out of sorts and probably in shock.

I managed to force open the Rover's door. They were both fine, if a little shaken up. The police and emergency services were on the scene pretty quickly and there were lots of flashing lights and urgency. An ambulance crew checked out the old couple. I don't remember how they left the scene, but I don't think it was in the ambulance.

In actual fact, my memory of everything immediately after the crash is rather fuzzy. I wasn't concussed, just strangely disconnected. When the police came, they marked out the scene and took over. I didn't have to do anything except answer a few questions, but even though I was found to be blameless, I remember feeling incredibly guilty, as though I had done something truly awful in being involved in an accident, even as the innocent party.

I was that disoriented that I didn't even think to phone the AA. Instead, the police called the local recovery people, who took my Bug away. I suppose I must have taken the train home from London, because I remember walking into the house and having to explain to Mum why I had appeared without the usual spluttering fanfare of the Beetle's air-cooled engine.

It took me a lot of hassle and a hefty storage fee to get my wrecked car back from the recovery people. Then, I had my first run-in with the car insurance system. I had a big fight with the insurance company while trying to prove to them the value of the car. I had done a few things to it, invested in the car to improve it, and it took nearly nine months to get the money I was due. I was lucky to have a really good insurance agent who put in a huge effort and got me more than I had actually paid for the car (although even that didn't seem nearly enough at the time).

I was keen to buy back the salvage as I was sure there was a way to rebuild the car. There wasn't. But I did later find a use for it on another project: a beach buggy. I even found a way to avenge my Bug some years later, when I modified another Rover 200, repeatedly, with a hammer ...

2.

YACHT YOU LOOKING AT?

Asking questions is a great way to quickly learn how things work, providing that the person you are asking knows their stuff. Even if they don't, you might still get a further insight or a lead to what or whom you can ask next; often you'll have to ask the same question many times, in different ways and of different people.

I've spent a lifetime gleaning all sorts of information that way. At first, I would drive my mum round the bend asking about anything and everything and then I'd do the same to my teachers at school and anyone I came across who I thought could teach me something. One of our neighbours, Will, and his mate Bob, were building a kit car, a Westfield Lotus 7 replica, on Will's drive and I used to hang around asking so many questions that they affectionately named me 'The Spirit of Ignorance'.

Most of my teachers weren't quite as amused though. I suppose I can understand that one child who won't just accept your pearls of wisdom but always asks 'why?', wanting to know more and taking your planned lesson in unpredictable directions, can get a bit wearisome.

In 1982, when I was 11, I went off to board at King Edward's School in Witley. My going to boarding school wasn't really so much a conscious decision as just the way things were done. My mum and her brother had both been to boarding school and my younger sister, Clare, and I would do the same.

As my mum was a widow, she managed to secure a small bursary but she still had to work three jobs as a nurse to cover the school fees. She worked at Guildford Hospital, was an agency nurse on weekends and during holidays, and did occasional admin work for the RIBA (Royal Institute of British Architects). I never once got a sense that she minded, it was all part of her work ethic, she always just wanted to do the best she could for us.

The day I was to start at my new school, Mum and I drove up to London, where all the parents and new students joined the governors and school staff for a service in St Bride's Church before going on to the Guildhall. One by one, the new students had to go into a large room where a number of serious-looking adults were sitting behind a wide, long table. We each had to walk across the room, stand on a little blue mat and answer some questions. The nerve-wracking ordeal was concluded with an 'I do' as we swore allegiance to the school.

Then there was just time for a quick goodbye with Mum, before being herded onto coaches and taken to the school campus in Witley, near Godalming in Surrey. It has always been a school of some sort ever since it was built in the nineteenth century – except for a period when the Royal Navy took it over during the Second World War. It's all very familiar to me now but when I first saw it, just like everyone else, I was impressed.

It was a bit daunting; everything was new and unfamiliar, and for the first three weeks we were not allowed to communicate with our parents, except by writing letters. It was the first time I had been away from home alone and I missed my mum, but the school were very good at keeping us so busy that we hardly had time to think about it and everyone else there was in the same boat.

There were a few tears at bedtime around the dorms but it helped us all to bond and forge friendships to last a lifetime. It may sound like it was a traumatic experience but I loved boarding school. It was just like having a huge sleep-over with all of your friends all of the time.

After the initial separation period, we were allowed out for our first 'exeat' – one day off campus. As we lived relatively close by, my mum could take me home for my favourite Sunday lunch – with lots of her legendary roast potatoes and lemon meringue pie for pudding – and get me back before bedtime.

One of the boys in my dorm, Steven, whose parents had not been able to come for him that day, was probably looking a bit forlorn, so my mum scooped him up and brought him home too. Steven and I are still close friends to this day. Though we didn't have much room, there was always an extra seat at the table or a spare bed at my mum's house and my friends were always welcome.

The rules at school were quite strict. Our beds were in dormitories – though beds would, apparently, have been the envy of pupils from the school's early years who had to wear naval uniform and sleep in hammocks! There was no talking after lights out and house masters prowled the corridors at night, cupping their ears against dorm doors to make sure we were all utterly silent.

That's always been a bit of a problem for me. I talk. It's one of the things I do best. The first time the dorm room door was flung open and the master's voice boomed out 'Who is making that noise?' I spoke up to take the blame, which was only fair because it was me who had been talking at that moment. Naturally, I wasn't the only one, but I was the only one brave, or foolish, enough to own up.

For my honesty, I was given the slipper. Each teacher had their instrument of discipline and I became acquainted with most of them, but the most feared was the housemaster's old trainer – called George – with a split sole giving extra bite when it connected with its target, my pyjama-clad posterior.

To avoid another meeting with George & co. I suppose I could have simply kept my mouth shut after lights out, but as that wasn't really very likely, I figured there must be a better solution. So, I rigged up a pressure pad under the carpet in the corridor that flashed a light in our dorm when someone came eavesdropping. Still, there were plenty of other things to get had up for, so George and I were never strangers.

There was a TV room in the common area, which we were allowed to watch at allotted times, but watching your show meant lots of negotiation with everyone in the house – or the use of brute force – so we would have much preferred to watch movies in our own dorm room. Except that wasn't allowed; watching TV was only allowed in the TV room.

The solution to this inconvenience presented itself a couple of years later, when I'd started doing computer studies. I was then allowed a computer in my room for devising and playing my own games, so, in order that we could watch a film in peace and quiet, I rigged a magnetic switch to some relays that flipped the monitor signal from video to computer the instant someone started to open the door.

Thinking back, we must have been a bit of a handful, but the school only had themselves to blame. It had always been a school that focused on technology and we were taught just enough to enable us to make trouble ...

In computer studies we were encouraged to devise all kinds of programs, including games. These were not the sort of lavishly produced games with movie-quality visuals that kids take for granted nowadays. They were really very basic.

In fact, BASIC (Beginner's All-purpose Symbolic Instruction Code) was the programming language we were taught, and my mate Aaron and I taught ourselves machine code so we could make our programmes a bit more sophisticated and capable. We were working with the latest, cutting edge personal computers – like the BBC Micro and the Commodore 64. The '64' stood for the whopping 64kb of memory that it boasted (which is about a tenth of the size of a poor quality picture on your smartphone).

Some of the coding techniques that we were messing about with might have been a bit risky; in particular, we often created self-writing code, mainly to save memory or speed up processes. Back then, this was a real programming no-no as even the tiniest of errors could easily cause the program to corrupt itself catastrophically with all kinds of random consequences. Aaron and I also devised what we called 'Annoyers', programs designed for the singular purpose of terrorising and confounding our tyrant of a computer studies teacher.

One of the perfectly reasonable rules in the computer room was that we were not allowed to play commercially available games. Rather un-reasonably, the computer studies teacher assumed all programs that were not written in BASIC must fall into that category, and that any game-like sounds must also be of contraband origin. This was in the days before Windows, so if you wanted to run several programs at once, you actually needed several computer terminals.

So, on one fateful day (for him), I was programming on one terminal while letting one of my other programs run on a different terminal. On the BBC Micro, sounds were generated by a number of variables in an 'Envelope'; my program randomly changed these parameters every few seconds to make all kinds of different sounds you might never think to program. Each time an interesting sound rung out I would leap over, push a key, and save that sound and its particular parameters. It was noisy and probably more annoying than listening to a fax machine through loudspeakers but Aaron and I considered it useful and entertaining research.

The teacher stormed into the computer room, no doubt incensed that we had such disregard for the rules and, without even questioning why no-one was sat at the computer making all the noise, wiped my 5¼ inch floppy disc that was sitting in the computer's drive. My program, the saved sounds and months of programming work, was all deleted in an instant. I was properly pissed off. So after that we set traps for him. Whenever he came to check the computer terminals for what the students had been up to, he would trigger our trap to be faced with rude words, an explosion of colour and sound, whatever we had come up with that day. One set of warnings flashed so suddenly and so loudly that he jumped back and fell backwards over a chair. Only teenage boys can laugh as hard as we laughed at that.

We were pushing boundaries and achieving fantastic results considering the technology we were working with, and we were right at the very beginning of the gaming age, on the fringe of computer greatness ... but, though my most excellent 'Dragrich' game might well have made it big (it was a dragon/ostrich which could punch and throw fireballs) my interests shifted and I went and did something completely different – there were just so many other things to discover and explore.

* * *

I had always enjoyed making things, and my A-level Craft Design and Technology course gave me the opportunity to create my first vehicle. It had no engine of any kind but it could go, quite literally, like the wind. It was a land yacht.

I had done a bit of wind surfing, which meant a lot of hanging around on beaches, and I had seen land yachts racing when the tide was way out. They looked cool, they were seriously quick and they cost nothing to run because the fuel was the wind, so I thought it would be a great laugh to design and build my own.

Designing and building, in my case, meant making use of whatever might be available to me in the school workshop. At the time, I wasn't really aware that you could just go to a shop to buy nuts and bolts and all manner of fixtures and fittings. We had bits and pieces in the workshop at school, but we never had quite what I needed, and so I had to come up with a multitude of engineering solutions to get round the fact that the parts I wanted to put my land yacht together were not sitting on the shelf waiting for me.

Well, they probably were sitting on a shelf somewhere, but I hadn't got a clue about that. With no internet, you couldn't simply look things up, and the concept of ordering things online for delivery the next morning belonged well in the future. So, I just made do with what was available to me, and made it work.

Because it is light and strong, the obvious material to use for my chassis was aluminium. There was plenty of aluminium tubing around in various sizes, but joining it all together wasn't as easy as you might think. My teachers told me that you couldn't weld aluminium, which isn't true. What they meant was that THEY couldn't weld aluminium. I was indignant when I later found out that people had been welding aluminium for years!

It was the same when our chemistry teacher taught us that you can produce hydrogen from water by super-heating the water to separate the H_2 from the O. I was fascinated, my mind rampant with all the possibilities, so I asked if there was any other way of splitting up these atoms and was told 'No'.

So, the following year, when the same teacher taught us about separating hydrogen and oxygen through electrolysis, I asked if this was a brand-new discovery. She obviously had to reply 'No'. Outraged, I asked her, 'Why wasn't I told about this before? What other things are you keeping from us?' It's fair to say that I was not a teacher's pet.

I was also regularly thrown out of our physics class for asking too many complicated questions that the teacher either couldn't or didn't want to answer. Within the first few moments of the class I had usually managed to put so many unwanted questions to him that I was banished to the library to find out for myself.

I'm not sure how many of those questions ever got answered but while I was in the library, I did research how to make my own dynamite. It's just a schoolboy's natural inclination to want to make something go bang – plus, I'd watched the two Dukes of Hazzard apply it to great effect on the end of an arrow. So, this became my new mission: making dynamite arrows.

Using a needle file and a photocopy of the school 'A-key', I made my own key so I could sneak into the chemistry lab to purloin the necessary ingredients. For a while, my excellent CDT teacher, Mr Bird, was concerned that I seemed slow during his class. What I was actually doing, was getting my projects done in half the time, so I could spend the rest of my time on the lathe turning my illicit arrow casings. It was maybe for the best that instead of successfully creating nitro-glycerine, I ended up with what I can only call nitro-caramel – it smelt great but didn't go bang.

If only they'd just answered my questions, I could have saved an awful lot of trouble. Still, it meant that I developed a healthy disregard for authority and it encouraged me to go on asking questions, whenever they pop into my head, and never accepting the first answer as the full story.

Anyway, the big project was still the land yacht and, unable to do any welding, I made knuckle joints and hinges, and screwed and bolted it all together. Manufactured parts and welding would have saved me time and weight but doing it my way became an exercise in finding alternative solutions.

We had a fantastic workshop teacher called Mr Bullock who had been a toolmaker in the Navy and he could make absolutely anything. If you asked him a question when you were trying to make something, his inevitable response was 'Look at the drawing, boy!' I still hear his voice in my head, with his rich Caribbean lilt, telling me to not be lazy, look properly; the answer will be there somewhere.

I learned so much from him not only about how to make things but how to work your way around a problem. Guys like him were amazing – spending so much time out at sea, maintaining engines aboard ships where there were no spare parts available, they regularly had to magic something up that would do the job and get them back to port. That kind of skill and ingenuity is to be admired and cherished.

With the chassis in hand, I needed something for the land yacht to run on and went with two motorbike wheels at the rear. Even I knew that those could be bought from a shop. They were tilted with a dramatic camber because I thought that would give me more stability, given how much sideways force there could be when the sail was full of wind.

On the front was a single plastic BMX Mag wheel which, legend had it, could get bent out of shape and would straighten itself back out if you left it in a freezer overnight. As most land yachts in those days were steered by hand, I had contrived a simple(ish) mechanism to allow me to steer the front wheel with my feet, leaving my hands free to deal with the sail and rigging. Part of the design brief I had defined was to utilise a standard windsurfing sail rig so that you could indulge in both sports without the extra expense and hassle of two sail rigs.

When it came to making the custom mounting which connected the mast to the land yacht chassis, I needed the coupling to have a thinner end to fit into the buggy and a thicker end that would fit snugly into the mast of the sail. All I had to work with were different sized tubes as there was no aluminium billet available in the workshop.

In any case, that would have meant a lot of machining (or a better design), so I decided that I could make the standard tubing thicker at one end by cramming some slightly thicker tubing on the outside. It had to be a tight fit, what is termed an interference fit, so I machined the inner tubing down so its outside diameter was fractionally under the inside diameter of the outer tube. It was a *very* tight fit. I actually bent the workshop press trying to squeeze it in. The trouble was, once it was part way in, I couldn't get it out again to skim a bit more off.

After a bit of head scratching, I decided that what I needed was a more powerful press and I reckoned there was one very nearby ... in the car park. My mum was visiting me, and I mounted my support on the bumper of her car and drove it, ever so slowly, into the workshop wall. The weight of the car and the power of the engine against the solid wall did the job.

The most vital component of the land yacht, of course, was the pilot, and my entire lanky frame had to be fitted in somehow. My idea was to create a kind of hammock made of transverse foam sausages suspended from two horizontal aluminium tubes, one at the shoulders and one under the thighs. I didn't know it then but it was a little like the fabulous Hyaline chair designed by Fabio Lenci.

The pilot would be completely supported whether sitting in a more upright position or fully reclined, and because the foam and webbing would conform to the shape of the pilot it would also provide lateral support during tight cornering manoeuvres. It was incredibly comfortable and I could lie in there for hours and hours. My mum spent ages sewing it all together – she was always amazingly supportive and keenly interested in everything that I did. She made me believe I could achieve anything and that every idea I had was pure genius.

As you get further into this book, you will notice that my mum crops up quite a lot, driving me here or there to pick up this car or that part, rescuing me from roadside wrecks and generally saving my bacon time after time. You won't notice my dad featuring so much; this is because he died when I was four years old.

Dad was an actual genius, a rocket scientist – well, more accurately a satellite scientist. He worked at the Royal Aircraft Establishment (RAE) in Farnborough and was involved in the Ariel satellite program and in particular Ariel 3, the first satellite to be built entirely in the UK. He was a very clever guy, but he also suffered from depression. It was never talked about in our family, but I know that he had some kind of breakdown because I remember Mum taking me to visit him in hospital. I was sent outside to play in the garden where there was a slide shaped like an elephant. Funny the things that stick in your mind.

In February 1976, three months before my fifth birthday, I was told that he had been in a railway accident. I remember standing on a chair looking out of the downstairs window waiting for him to come home, but he never did. It was eventually explained to me, by my mother and Fran and Jean, my twin aunts, that he would never be coming home. I cried for what seemed like weeks.

It was much later that I learned that he had actually taken his own life. I spent much of my young life confused and eventually angry that he left us like that. How could he have abandoned us in that way? In my teens I sometimes fantasised that he had been abducted because of the work he was doing at the RAE. Was it top secret? Was he still working under duress for some devious foreign power in a secret underground lab somewhere? I guess anything was better than that feeling of abandonment.

I've never felt I reached his level of brilliance or achievement. Maybe that's been a component of my constant drive to attempt things just out of reach. I did pause for a moment when I eventually made it past his age.

I have worked hard to forgive him. He was just a human being, struggling to make sense of the world around him and within him, and somehow, he couldn't make the two add up. It must have been unbearable for him. It was unbearable for us too, especially for Mum, but she stepped up and did double duty, supported by friends and family and she gave us a wonderful childhood. We never wanted for anything, especially not love and encouragement.

Whenever I had free time, the school would let me go out and test my land yacht around the grounds. We had lots of playing fields, service roads and a big expanse of tarmac – loads of open space where I was unlikely to do much damage to anything except the land yacht or, possibly, myself. It didn't go so well on soft ground but on the tarmac and firm grass it really flew.

When it wasn't windy, we would take the mast out and hurtle it around as a gravity racer – like a skeleton soap box cart. The school grounds were on different levels and I could coast down slopes all the way from the top end of the school to the bottom end, picking up speed all the way. It gave me a good appreciation of how useful brakes are, because this thing didn't have any!

On a beach or open ground, you don't really need brakes so much because you are on the flat and you can use the sail to control your speed. Piling down a slope through the school without a sail was another matter. It could so easily have ended in disaster but somehow, we survived.

I say 'we' because the land yacht was attracting quite a bit of attention. Obviously, my mates all wanted to have a go, and anyone was welcome, but as it was built to fit my lanky frame, very few of them could. My shortest friends, Sue and Alison, then worked out that they could take a spin by talking me into letting them sit on my lap while I manoeuvred the craft down the hill. All of a sudden, girls I'd never even talked to came up and wanted to befriend me, hoping to get a ride down the hill.

I know what you're thinking. This had been my ulterior motive for building the land yacht all along: get a few girls interested, snuggle up in a custom-made hammock and sail around the playing fields in a sparkling, hormone-dusted cloud of teenage lust. Nothing could be further from the truth. No, this was a speed machine, not a love machine and I was, in any case, desperately ill-prepared to cope with that sort of attention from girls.

Girls to me were just mates and I was always the awkward, gawky one at the school disco, way taller than anyone else, a flag-pole on a parade ground. I always felt conspicuous wherever I went, and I was so hideously lacking in confidence that dancing was definitely out of the question. I just knew I would look like a deranged giraffe on the dancefloor, so that was best avoided. Instead, I distracted myself with other pursuits.

The land yacht was definitely one of the more unique projects that came out of my CDT class and it proved to me that being slightly different could get you noticed in a good way. The way that it attracted girls was an unexpected side effect that I didn't really know what to do with, but the notion that I could use a contraption like that to disguise my basic shyness was one that I banked for future use.

* * *

Some things turn out to be assets that keep repaying your initial investment over and over again – a gift that keeps on giving, as they say – and so it was with the land yacht. I played with it for a few years afterwards, including going with a couple of friends to Wisley Airfield, near Ockham in Surrey, where we took it up to silly speeds, although not under sail power.

Wisley was one of many Second World War air bases scattered around the south-east of England. Most of them were fighter bases protecting London and the south coast but Wisley was a bit different. During the war, the Vickers aircraft that were built nearby at Brooklands and Cobham were shipped here by road to be tested before being delivered to their units.

They even had a Wellington bomber with a difference here. Instead of a rear gun turret in the tail, it had a jet engine. Because Vickers (and subsequently BAC) were testing aircraft here after the war right up to 1972, a concrete runway (as opposed to a grass strip used previously) was laid in the 1950s. In the modern era, the runway was too short for large aircraft to use, so the powers-that-be pretty much abandoned it, leaving it for people like us to sneak in and do our own thing.

What they didn't want, of course, were the wrong kind of people – I mean even more wrong than us – using the runway. To prevent smugglers doing midnight drug runs they installed staggered concrete bollards and Armco barriers to stop aircraft landing. Needless to say, those didn't stop us; all they did was provide us with an extra challenge as we slalomed between the gaps.

I had an old VW pick-up truck by then that we would drive along the runway. I would 'surf' the land yacht behind it on a tow rope, without the sail, just to see how fast it would go. You could go faster and faster by whizzing out to the edge of the tow rope like a water skier behind a power boat, but then you had to dodge the bollards. Fantastic fun.

It was a real buzz thundering down the runway, the old VW screaming through its gears up ahead, my ears filled with the sound of the land yacht's bike wheels humming on the rough concrete and bucking over the ancient, weathered expansion joints. Despite the jarring and clattering, I never doubted that it would all hold together. Well, never for long, anyway.

We were all more interested in how fast we could go. With no power of its own, the land yacht was clearly limited by the speed that the VW could do, but we made it to nearly 100 mph which was very satisfying, a bit scary and quite possibly insane.

For a while I joined a land yachting club and hauled my machine around to a few meetings. There was another old air base somewhere farther north where land yachts of all shapes, sizes and colours would come together and race. I wasn't as fast as most of the enthusiasts who were mainly using professionally manufactured rigs, but I had the satisfaction of knowing that mine was all my own work, and probably had the comfiest seat, thanks to Mum.

I also took it to various beaches with family and friends. We visited Gwithian, near Hayle beach in Cornwall, which has a fantastic wide stretch of sand when the tide is out. At three miles long, there's plenty of room for a land yacht to do what it is meant to do – carefree sailing, just short of the sea.

The problem with sand, as anyone who has ever tried to drive fast on it will tell you, is that it's rarely consistent. Hard, flat sand is brilliant, but soft sand or damp sand does weird things; your wheels dig in, forcing you left or right, losing grip or robbing you of power. In a beach buggy, that's not such a big deal because you have good wide tyres and enough power to get you out of trouble. In a land yacht like mine, with skinny wheels, it's completely different.

On hard sand you zip along with the wind – quite literally – in your sails, in your hair and up your trouser legs. When you hit a patch of soft or wet sand and a wheel digs in, anything can happen. You can make a miraculous and skilful recovery to keep yourself on an even keel, or you can flip over, up in the air, into a sand bank, or all three. Before you know it, you are wearing your sail as a poncho and wondering if you'll ever get all those grains of sand out of your various orifices.

I had a bit of a mishap one day when the 'interference fit' tubing failed. There was a good breeze blowing and I'd been scorching across the sand, but years of stress and strain caused it to crack. My first thought was that we were going to miss out on playing in a perfect breeze for the rest of the week.

However, all was not lost. Along Penzance Harbour we found an engineer who used his skill, his experience, his lathe and a lump of rusty metal to fabricate exactly what I needed to replace the broken fixing. Guys like that are worth their weight in gold. He turned that lump of metal into a perfect fitting in what seemed like next to no time, meaning we could get back to messing about on the beach.

The land yacht gave me years of fun, but I think that it was probably 'repurposed' at some point when parts were needed for another project – although I'm pretty sure I still have some bits of it somewhere. Nothing useful, after all, should ever be thrown away, and the land yacht turned out to be incredibly useful in many, many ways.

* * *

A few years later, when I did my first final-year degree project – I had more 'final' years than are strictly required – I had to submit three different ideas for a design project that I could actually build.

Having been asked for 3 ideas, I came up with about 20 – pinging out ideas has never been any trouble for me. First,

I wanted to build a car, but they wouldn't let me (I ended up building one anyway, a Fugitive sand rail, but it did nothing for my degree).

Then I had an ecologically sound idea for saving water when taking a long shower (I love long showers). My shower design would clean and recycle the water that you used during each shower, effectively allowing you to luxuriate in the same water needed for a 30-second shower for as long as you want. For some reason that wasn't suitable either. Too tricky to get right, is what I think the course tutors said.

Finally, it was agreed that I would concentrate on a car trailer that would also work as a lift to raise the car so that you could work underneath it. Whether you were tackling a mechanical issue on your driveway, or perhaps you were an AA guy rescuing people at the side of the road, this trailer would give you a facility that is otherwise only available in a garage workshop.

If you have a trailer that you use from time to time, you will know that at all other times the trailer can be a real pain in the neck because when you are not using it, it just takes up space. This trailer would have worked as a car transporter, a car lift and could even have provided an extra parking space with one car on top and another slotted in underneath.

It was coming along really nicely until I hit a problem that I just couldn't seem to solve. I had come up with a design for the wheel configuration on the trailer which meant that the front wheels needed to be able to steer to follow the car around corners. This meant that when you were reversing, they would turn in the opposite direction and lock the whole thing up (just like when you reverse a car and let go of the steering wheel), so I needed a mechanism that would flip the caster angle of the wheels when reversing.

I tried all sorts of solutions to make it work but just couldn't figure it out. It was really annoying, I knew that there was a simple solution, I just couldn't see it yet. I wasn't too worried. The reversing problem didn't mean that the basic idea wouldn't work, and a solution to going backwards would present itself eventually.

However, my lecturers weren't quite as sanguine about this impasse and were getting nervous that it wouldn't be resolved in time for the final-year show. So, they persuaded me to take a second 'final year' (with a different project) and go for something simpler, and they particularly liked my idea of a Recreational Amphibious Wind-powered vehicle (RAWv).

Actually, that change of direction was quite a shrewd move, because all of my messing around in the land yacht had provided a wealth of research to help me design something that was a lot of fun. I recycled some of my design from the land yacht and, for one of the prototypes, adapted a windsurfing board, cutting it up to make the body of the vehicle.

The final design centred around a stainless-steel frame, which provided the structure for a mini RIB (rigid inflatable boat) style hull which could be deflated for transport and storage. Keeping with the portable theme, I ditched the windsurfing rig in favour of a power kite. The idea was to create a buggy that could race up and down the beach but then splash straight into the sea and just keep on going.

The clock was ticking, and there were now only a couple of weeks before the final year show. I had to make a video of the RAWv in action to prove the concept and then have something to explain in my (still unwritten) dissertation. Having done all of the research and built a prototype, to finish my final, 'final year' I needed evidence that it did actually work.

My best mate at uni, Dave Davenport, was to act as cameraman. We drove down to the beach at West Wittering in Sussex late in the day, with the tide leaving enough beach for me to drive the RAWv on shore, but deep enough water for me to launch into the sea and sail away. We also didn't want to have too many people around, getting in the way, but we had to get it all done before we lost the light.

We had a generator in a van parked in the dunes with the longest extension lead you have ever seen stretching all the way down to Dave, who had the surplus gaffer-taped around his waist and was standing in the water in order to get the best shots.

Dave was a brave man – it was an electrocution waiting to happen. To our delight, even though my prototype was rough, it did what it was supposed to do and proved capable of operating on land and in water. I don't think Dave has ever been so happy to leave a beach.

The important thing was that it got me through the practical part of my degree. Having completed my design and proved that the prototype worked, I could now write the thesis. I had just over a week! I cranked up the Prodigy, and started to hammer the keyboard, printing out pages and sticking in photos as I worked through days and nights.

I finished just in time, set up my display at our show, ready to be marked. I managed to walk away with a 'Desmond', a 2:2, otherwise generously known as a 'Gentleman's Degree' – an accolade usually reserved for bright slackers or those easily distracted from their studies. But I was qualified, and ready for the next adventure ...

* * *

Several months later, Dave had got a job working for PDD, a design consultancy in London, and I was travelling up to join him and a few others for a night out. As I was bombing along the M3, late as usual, putting the world to rights with Dave on the phone, the solution to that trailer-lift reversing problem suddenly popped into my head.

There's an electrical reversing signal going through the caravan socket to the trailer lights every time the towing vehicle is put in reverse! So, all I had to do was to tap into that signal to energise a solenoid which would release a pin allowing the caster to flip under the force of the reversing action. Now that I had the solution to the problem, I could finally crack on and build that prototype trailer.

Well, that was 20 years ago, although it seems like only yesterday, and I still haven't quite got round to it. Yet. I suppose that they have been a fairly busy couple of decades. But one day soon, I might still get around to building that multi-purpose trailer ...

3.

VAN-DERLUST

I have always been a bit partial to a van. I really like their versatility and the fact they're useful in so many different ways. You can use them to get from A to B, fit loads of stuff in them, lug around engines, axles, surf boards, gangs of mates – you name it. Or you can sleep, or even live in them.

This interest in vans began, rather randomly, at school. In my senior year at school, the teachers would host little evening receptions for students in their homes on campus. It was meant to be a way of introducing us to grown-up socialising. We would be assigned to groups, to ensure that we didn't just huddle with our mates mumbling as usual, but instead developed the art of conversing politely with people we didn't really know very well. (Occasionally, we were even allowed a glass of wine.)

At one of these sessions, we were talking about 'gap year' plans, and movies, and the Cliff Richard film *Summer Holiday* cropped up. Everybody in my discussion group felt that driving a bus around Europe sounded like a real giggle. They all agreed it was a good idea, and the subject was forgotten ...

Except by me. I didn't think it was a good idea – I thought it was a *brilliant* one. And I filed it away for future use.

I had always loved the anticipation of journeys. My dad had a Mini Clubman estate – the one with the sliding windows at the back – and as a kid, my family used it to travel down to the West Country in the summer to spend time with grandparents, aunts and godparents.

I can still remember it once being so foggy as we drove along the A303 near Stonehenge that we could barely see the end of the Mini's short bonnet. It was like driving in a cloud. Dad crawled along until he managed to pull the car into a petrol station and we had breakfast at a Little Chef waiting for the fog to clear. I was beside myself with excitement. The journey had barely begun and we'd already had a thrilling adventure. That Little Chef has morphed into a Starbucks now, but I always get a twinge of nostalgia when I pass it today ...

When I left school and enrolled at South Bank University in London it was deemed by the college authorities that I lived too close to qualify for student accommodation. I therefore had to travel into uni by train. The route I took from Waterloo station to the college took me past an old church that looked as if it had been derelict for years and years. I noticed that there was a little passageway through to a car park at the side of the property and an idea started to form.

Maybe I could get hold of a camper van and live in that church yard, thus saving myself a fortune in rail fares and time? Even better, if a camper van would fit in there, might there be room for a double-decker bus? I thought there just might be.

Fired with enthusiasm, I got in touch with the church's owners. They were amenable and very readily agreed that I could rent a space big enough for a bus and my car, for less than it would cost each week to park my car near the uni. They even put it in writing.

My plan was fast beginning to hatch. Now all that I had to do was to buy a bus ...

I would have loved to get a Routemaster, the most iconic and recognisable of all London double-deckers. Specifically, I wanted the extended wheelbase version, designed to accommodate eight extra seats, which is recognisable by its extra little square window in the middle.

The Routemasters were a masterpiece of design and engineering. They had a revolutionary aluminium monocoque body, meaning they were very strong and light, so fuel consumption was kept low and they didn't rust. All of this meant that they were greatly valued for second-life applications once Transport for London sold them off after active service. They were bought by tour operators and local councils around the world, who could get more use out of them. So, unfortunately for me at the time, very popular meant very pricey, and my meagre savings and student loan were just not going to stretch to one.

This was a particular shame as the rounded shape made them look like an old-fashioned toaster, and my mate Dave and I fantasised briefly that we could polish up the aluminium and attach something that looked like two slices of toast sticking out of the roof and serve toast to hungry students. It may not have been a *Dragons' Den*-impressing billon-pound business idea but it would have looked pretty cool!

Apparently, the next best thing to a Routemaster would have been the Bristol VR, a more modern, square-shaped bus with closing doors at the front and the legendarily reliable and very desirable Gardner diesel engine at the rear. However, if I was going to have a modern-looking bus and not the Routemaster, I wanted one with more space inside. Also, the Gardner engine came at a premium, but as I wasn't going to be doing a lot of miles I didn't need to stump up for that. In the end, the bus I opted for was the mighty Leyland Atlantean.

I found one in a bus depot on the A13 in Essex. However, the very day that I bought it, the church sold its derelict property to a property developer who would be turning it into flats, and my parking place was gone. Bugger!

I hadn't actually picked up the bus from the depot at that stage and I went back to Essex to explain the situation. The guys at the depot wouldn't call off the whole deal but they had an alternative solution. They offered to let me keep the bus at the depot for nothing.

It was a good offer. They had acres of space and, more importantly, all the facilities you could dream of for working on a bus. Huge heated workshops with lifts, guillotines, presses and folders – and all there for me to use whenever I wanted.

Even though I wasn't right next to the uni, I was closer to town, I had no rent to pay ... and living at the bus depot had the added benefit of getting me free rides in and out of town on the tour buses which operated from the depot.

So, it seemed I was to keep the bus, and if I was to keep it, I had to learn to drive it. When I bought it, the bus had just come out of service in the Ribble Valley in Lancashire, with seating for 70 passengers. In order for me to be able to drive it without a bus driver's PCV (Passenger Carrying Vehicle) licence, I had to get the Atlantean de-registered, meaning that all but eight seats had to go. So, with a little help from my friends wielding angle grinders and pry bars, the Atlantean was soon ready to be examined. Quite remarkably, it was re-classified as a car, meaning that I could drive it on an ordinary licence.

However, being *allowed* to drive something and being *able* to drive something are very different matters.

I had a short lesson from one of the bus drivers at the yard, who took me out around the block. The Atlantean was surprisingly easy to drive, thanks to having air-assisted everything. However, it took me a while to get used to the fact that it was many times the size of my Beetle, and because you sit so far in front of the front axle, you have to go a car-length further into a junction before turning a corner.

Now that I could take the bus out on the road, I drove it home and parked it on the street outside Mum's house, much to the obvious consternation of the neighbours. Luckily for them, I only stayed about a week before I found a more suitable temporary parking spot, and then I moved it out to Essex. It was fun to show off my latest folly and very handy to have easy access to mains power and my tools to get the project started.

As soon as I got it home, I took Mum out for a spin in the bus. It's a bit terrifying when you're doing the limit (30mph), you've got one wheel right up against the gutter and the other on the white line in the middle of the road, the bollards on a traffic island are coming towards you, and there's a queue of people behind you getting impatient. I soon found my confidence though and would hop in the bus to pop down to the local shops for a pint of milk – just because I could.

I had not forgotten the original inspiration for getting a bus and my mates and I decided we had to take it on a road trip. Maybe not all the way around Europe because, well, we didn't have all summer, we were all skint or working summer jobs, and the bus was now just a big, empty shell with eight seats and nothing else in it. Still, we set off on a few local adventures – daytrips to the beach and a return trip to our old school campus where it had all started.

After all that messing about, I had to really get cracking on my conversion if the Atlantean was going to be habitable by the start of the next uni term. It had its doors at the front and the staircase just behind the driver, which left the rest of the space inside the bus pretty much unencumbered for me to work out my design as I pleased.

I had left the rear seats on the lower deck in place and the rear wheel arches were the perfect foundation for a very tidy galley style kitchen. I managed to squeeze in a gas four-hob cooker and grill, a water heater, some cupboard space, a fridge, a microwave and even the kitchen sink. The lower deck was slightly taller than the upper, so the shower needed to be downstairs; unfortunately, though, there was only enough headroom for me to stand up straight in the stairwell, so for maximum luxury in the shower I elected to steal a bit of headroom from the upper deck. The subsequent box jutting into the upstairs floorspace was the perfect position for a TV pedestal.

In fact, plumbing dictated much of the layout. Having committed to the shower's position, to save space I added a caravan toilet and sink to the same cubicle to form a kind of wet room. I even had a special nook to keep the toilet paper dry. Then I needed to think about water.

I needed three water tanks: a freshwater tank, which I ended up hiding in a steel reinforced bathroom wall; a grey-water tank which took used water from the sinks and shower to flush the toilet; and a black-water tank for them to flush into. To keep the bus stable on the road, the tanks needed to be positioned in or near to the floor – water is heavy and if it's sloshing about above the centre of gravity it will seriously affect the handling and stability. Of course, pretty much everything else needed to operate the bus is also largely positioned in the floor – cables, batteries, pipes, the automatic greasing system – so the tanks would need to be custom-built and fitted. Luckily, I found a company in the Midlands which specialised in that.

It was time for another road trip, this time with Mum leading the way in her Nissan Bluebird. This was in the days before sat nav, so Mum had plotted the journey using paper maps, writing down detailed instructions for herself to refer to en route, allowing me to concentrate on driving the bus.

The journey did not go well. Apart from the fact that the Atlantean couldn't get above 40mph, even when giving it the beans on the motorway, and we had to keep pulling over to let people pass in every layby on the single carriageways, Mum got us hopelessly lost in a housing estate and I really got to practise my reversing and 30-point-turn skills. On the upside, I became intimately familiar with the residential streets and cul-de-sacs around Kenilworth.

Tanks fitted, I had to remove some windows and replace them with aluminium panels to protect my modesty in the shower. Happily, everything I needed was at the bus yard: the tools, the materials and the know-how, courtesy of some of the old boys there. I just had to provide the labour. To replace those windows, I had to first drill out 2,500 pop rivets and then attach the aluminium panels with exactly the same amount, using a manual rivet gun. It was worth it; it gave me my privacy and the bus looked really smart, but I couldn't use my hands for several days afterwards – I'd even ended up having to use my wrists to operate the rivet gun towards the end of the three-day job.

Now that the shell was complete, and all the underlying systems in place, I could start to create my double decker dormitory/palace on wheels. I built a huge bed on the top deck, covering the entire width of the bus above the driver's area, with curtains round for privacy and storage underneath. The rest of the upstairs was a lounge area and my study with my huge drawing table and a plan chest.

Downstairs I made a sound-proof cupboard for a generator and installed a gas heater with blown air ducted around the rest of the bus – it took only about ten minutes to heat up the bus on a cold day. Although, that was if the butane tank hadn't frozen, which it often did. An upgrade to propane sorted that out. I even had a space to store my bicycle on the luggage rack opposite the stairs. It was all really starting to come together. It wasn't quite finished by the start of term, but I still moved into my new student digs and continued to work on it in my spare time.

The 5kw generator gave me 240v electrics, and after a while I got to thinking about how crazy and really inefficient it was that I had to run this big generator just to get the small amount of power needed for the odd light and the sink pump, so I installed some batteries. It meant that I could just run the generator to charge the batteries and power the low-voltage appliances from the battery pack, rather than having the generator running all the time.

Of course, this was way before companies like Honda started doing battery packs with inverters, or their own generators, and it started me thinking about how we might solve those sorts of electrical problems on other vehicles. Those are thoughts that stay with me until this very day.

My bus was a really cool place to be and for a while my plan worked a treat. I got free rides on the buses in and out of town, I could study on the journey and they dropped me off right by university. The buses ran late at night, so I could do all my socialising, or go to the gym or whatever before returning home. I had a job at Pizza Hut near the depot and on work nights I would cycle back to my Atlantean with pizza tucked inside my jacket, the wafts of warm pizza aroma giving me an incentive to step on the pedals and get back to the bus. On most weekends I went home to my mum's, returning with clean clothes and treats. And then ... winter came.

This meant that some nights the temperature dropped to about -7°C, and if the gas bottle had frozen, sleeping in the Atlantean was really grim. This was not ideal RV living. I was shivering in my bed even under a duvet, in my sleeping bag, fully clothed and wearing a balaclava. The condensation running down the walls made everything damp, and at that point Mum persuaded me – well, kind of insisted, really – that I had to come home. That move may have saved me from terminal pneumonia but it was to spell the beginning of the end for my bus.

That meant I had to get the bus home from Rainham. With no London Congestion Charge or Low Emissions Zone to worry about in those days, I usually went home the direct route through central London, and I was enjoying the architecture around the Science Museum and the Natural History Museum in Kensington when some pillock cut me up. I had a bit of a wobble trying to avoid rear-ending him and an overhanging tree branch smacked into the front of the bus's top deck.

It buckled the aluminium, crushed the fibreglass and smashed the front and side screens. It was a blow, but the bodywork on the top deck is not integral to the structure of the Atlantean and the bus was still safe to drive. Having had a quick look, I was fairly confident that I would be able to put right the damage without too much trouble.

When I arrived home with my sadly battered bus, I decided that the best place to park it temporarily was in the council lorry park. Pizza Hut had been brilliant and offered me a job closer to my mum's house, so I went to bed that night reasonably content, happy that I would soon be able to get cracking on the conversion again and there was still the possibility of that European summer adventure. Besides, I was sleeping in my own, comfortable bed, in a heated house, so at least I could dispense with the balaclava.

Unfortunately, when I moved back home there were several families of ne'er-do-wells staying in the local area. They had been causing general mayhem and they paid a visit to the lorry park late one night. First, they drained the diesel from all of the lorries and other vehicles parked there, then smashed a few windows to amuse themselves.

They must have then realised that my Atlantean wasn't just your average bus and took a look inside. To them, it must have been an Aladdin's cave. They stripped all of the fixtures and fittings from the generator to the shower, even the kitchen sink, taking anything that they could easily remove and ripping out any new metal that they could sell for scrap ... and one of them then took a dump in the now non-functioning toilet.

That was not a pleasant thing to find the next day. As luck would have it, though, they came back the following night and removed the toilet as well, with their turd still in it. Clearly, they wanted to keep their business their business.

I was heartbroken to lose everything from the bus (apart from the turd, which I was quite glad to see the back of). When I started making a tally of the damage that had been done, and the kit that had been stolen, then factored in what it would cost to repair what the Kensington tree had done upstairs, it was clear that my meagre student budget was never going to cover it. It was a disaster.

It was also upsetting that the police seemed powerless to do anything, even though they knew exactly who was responsible. They said the chances of securing a conviction were practically nil. The local council were even less interested. The miscreants moved on, as they are wont to do, to blight some other neighbourhood, at which point the local council leapt into action and condemned the Atlantean as a health-and-safety hazard. I had no choice but to sell my poor old bus for scrap. I think the bloke who came to take it away actually took it back up north, so I like to think that the bus went home to die.

* * *

My escapades with the Leyland Atlantean didn't deter me from the idea of building or converting a vehicle that I could live in rather than commuting into town every day. So, I bought an old VW T2 Camper, an army van that had done its national service with the Dutch armed forces. It was pretty cheap, mainly because it was a fairly unattractive drab green, and a left-hand drive, so nobody else wanted it. The interior was stripped bare, except for the front seats, but it was rust free and the engine was good.

I created a comprehensive fold-flat interior that allowed me to haul around big bulky loads but also unfolded to give me a full width bench seat, a table, a bed, a cooker, a fridge, sink and other fundamentals. The owners of a recording studio right next to the student union building let me park it in their car park every night as long as I agreed to act as a sort of night guard, letting bands in or locking up after late-night recording sessions.

I stayed in the camper during the week for the best part of a couple of terms, driving the van up to town on Sunday evenings having spent the weekend at home. On misty nights, the drive was a bit of a challenge because the heater took so long to warm up that I was usually at my destination, having driven there with the window wound down, before it started to clear the fog on the inside of the windscreen.

Eventually, I saw sense and just rented a flat with some uni mates, but it was in this camper that I finally got around to taking that European road trip, my first taste of a camper road trip since I was a child.

My early childhood trip had been an amazing adventure. When I was six, Mum took Clare and I on a trip to the United States. For some reason, I had convinced myself that America would be all jungle, like in *Tarzan* on TV, so it was quite a revelation to look down at New York City through the glass observation floor in the World Trade Center's Twin Towers.

Our epic trip took 11 weeks and we covered a lot of ground. We spent Thanksgiving with our friends, the Bowers, in Washington, and then travelled in their RV, down through Virginia to Florida, where we picked oranges, saw alligators in roadside ditches, watched Shamu at Sea World and visited the magic castle in Disney World.

The RV was huge and I loved being away in it. Mum and Clare slept on a bed in the living room area and I was given an excellent 'bed' in a cupboard above them. I remember saving some extra bed time cookies in the top pocket of my pyjamas. I fell asleep only to wake up in the middle of the night with itchy cookie crumbs everywhere. I brushed them out of my bed onto Mum and Clare while they slept down below. I'm sure I got away with it.

It was a proper road trip taking in so many sights: a chocolate factory with live humming birds, the rockets at Cape Canaveral, USS *Nimitz*. We even saw President Jimmy Carter get in his helicopter. We had an amazing time, seeing so much of East Coast America.

So, as a student at the grand age of 21, I decided that it was time that I set out on another road trip. I was handily equipped with a European left-hand drive VW Camper, and my mate Steven and I had our sights set on the Expo '92.

The event was hosted by the city of Seville in southern Spain, with over 100 countries contributing to a massive spectacle on the theme of *The Age of Discovery*. Between April and October of that year, almost 42 million people visited Expo '92. Steven and I were not among them.

We excitedly set off a couple of weeks before the end of the summer holidays, thinking we could just squeeze in a few days pottering around Seville before the start of term.

To save our money for Seville, we decided to avoid the toll routes, which meant keeping to local roads all the way. What could possibly go wrong? In preparation for the trip I had even bolted a little compass in a bubble onto the dashboard.

It felt like we were driving through France forever. As we finally got to Paris we also decided to avoid the Périphérique, as on the map it looked like a toll road. We almost instantly got lost as we hit the suburbs. The local signage made no sense to us and only made things worse. After hours meandering through local traffic we finally made it to the city centre.

We found ourselves driving around and around the ridiculous multi-lane roundabout that is the Arc de Triomphe. After the second or third time around, it was time for some decisive navigation. We took a compass bearing south-south-west and headed for Bordeaux, following the compass all the way out of town. It had only taken us four hours to get through Paris.

It had been dark for hours by the time we were close to the Pyrenees and the border with Spain. Then, as we approached the foothills, the heavens opened and we were battered by torrential rain. Just as I was starting to wish that I could see where we were going a bit better, we rounded a bend and I could immediately see exactly where we were going ... and it wasn't good.

The entire road was flooded by a small lake. There was no telling how deep it was, and there was no time to stop before we ploughed into it and felt the van aquaplane. I had never known anything quite like it before. We were power sliding sideways across the water for what seemed like an age with no control over the steering or brakes whatsoever!

Dramatic images flashed through my panicking mind of us hitting a verge, flipping over, tumbling down an embankment and bursting into flames. However, just as we were about to plummet off the road into a big ditch, the aquaplane drift stopped as suddenly as it had begun, the wheels gripped solid ground and we were on our way again. It was a very lucky escape, not to mention a dramatic introduction to the dangers of standing water and aquaplaning. We pressed on through the mountains and on into Spain.

It was now the early hours of the morning. I was struggling to stay awake, Steven was snoring in the passenger seat, we were freezing cold and about 100km off Madrid. I woke Steven for a chat. We worked out that if we kept going, we could get to Seville, see all there was to see and head back to reach the Channel just in time to catch our ferry, but only if we stayed awake for the next three days and drove like crazy. We agreed that wouldn't really be much fun and decided to abandon our Seville mission in favour of taking the scenic route home instead.

We turned back and parked up in the first deserted lay-by we found. It was pitch-dark outside, and definitely time to sleep.

The next day we woke up and flung open the side door. Closing our eyes against the searing sunlight that flooded in, we yawned and stretched, two scruffy, almost-naked students scratching our balls and looking out at ... well, I suddenly started getting flashbacks to *The Life of Brian*.

All around us were scores of families eating al fresco lunch and staring at us with expressions somewhere between horror and disgust. We slammed the van door shut again. Peeking out of a window, we realised that we hadn't stopped in a disused lay-by – we had parked in a local picnic area and beauty spot, with kids running around and food laid out neatly on tables.

Dressing quickly, we stumbled out into the fresh air and down to a nearby waterfall to have a rudimentary wash before we headed back into France. Steven screamed so loudly when he put his head into the freezing mountain water that the poor families probably thought I was murdering him. I suspect they were very happy when we climbed back into our old van and disappeared.

We made fairly good time as we journeyed back towards France ... until we got a puncture. As we were staring at it, a police car pulled up. The cops were amazed when we shrugged, found our trolley jack and changed the wheel in 10 minutes flat. They said most locals would have abandoned the van, thumbed a lift to civilisation, and returned a day or two later with a spare wheel.

We continued on our way, but the camper van wasn't finished with us yet. The next day, we crested a hill with a long straight descent and beautiful panoramic view in front of us. Suddenly the van started sputtering and then cut out completely. No engine, no electrics. We coasted down the hill as, in the distance, I could see a motorway services at the bottom.

We managed to roll all the way onto the forecourt but unfortunately the services were closed, or, more accurately, were yet to be opened. It was still a construction site and not a soul was around. I opened up the engine bay and had a rummage around. After quite a bit of further rummaging I discovered that somehow the wiring loom running behind the engine had worked its way loose and been sucked into the cooling fan. This had chewed the wiring to pieces and made everything cut out. No wonder we had lost our electrics. We had tools but needed to get our hands on some new wire.

On the way down the hill we had passed a second cluster of buildings on the other side of the carriageway: the other half of the service station, also under construction. It was worth a look; I weaved my way over, dodging the traffic like a giraffe playing *Frogger*. I was in luck: a bunch of Spanish electricians were busy with the first fix. Unfortunately, none of them spoke English and I had no Spanish.

I used my best sign language, mime and gurning to blag some of their cable. It took a while, but I was able to strip the cable apart to patch the wiring loom back together and we were eventually back on the road.

By the time we got to northern France, Steven and I felt confident that we would reach Calais in time for our ferry. We had been given dire warnings about the delays and the excruciating expense involved if you didn't show up in time for the ferry on which you were booked. There was no Eurotunnel back then. The ferry companies ruled the roost.

Our confidence evaporated as soon as we began losing a bit of power and a disturbing rattling started coming from the engine. Initially, we did what any sensible person would do – we pretended it wasn't happening and pressed on. However, we lost more and more power and just couldn't maintain our speed. We had to stop to take a look.

The dual carbs had started to shake themselves off the engine. We were close to Calais, but we had so little time. If I spent time to do a repair, we risked missing the ferry, if we limped on we might not make it at all. I got busy and managed to lash the carbs back into position.

We made it to the port ... just in time to see the ferry disappearing into the distance.

Disaster! Steven and I were devastated. What could we do? We nervously explained what had happened to one of the ferry officials. He shook his head, then smiled: 'Not a problem, lads. Park over there and we'll get you on the next one.' He didn't even ask us to buy another ticket.

Despite its appearance, my scabby-chic bay window van was very comfortable and extremely practical. I suppose if I had kept it for long enough, I may have got around to painting it a less drab colour. The pearlescent 'tangerine orange metallic' (paint code: YR563M) paint I used on the 1979 VW panel van we restored on the show would have been just the ticket but, back then, I was more interested in the utility and performance of my workhorses than their aesthetics.

In the end, my sister Clare's, need was greater than mine, so I swapped it for her first car, a slammed lilac 1965 Beetle that I prepared for her eighteenth birthday. Meanwhile, I 'upgraded' to the slightly less desirable, but faster and more comfortable VW Type 25 'wedge' van.

They were squarer in style but handled really nicely and had much more room both in the cab and in the rear. The early T25s had air-cooled VW engines, but later ones used water-cooled motors – a bit smoother and quieter, and still right at the back of the van.

Type 25s used to come up for sale quite regularly and I bought one or two (OK, several) in quick succession. That was when I started playing around with the engines. It started off with a stock van that I bought from a dodgy car dealer. The engine had blown and the head studs were broken. I tried to repair it but it was a goner.

So, the engine was scrap but what I really liked about that van was that it still had the French polisher's signwriting on the side. I figured it would be really cool to go burbling around in what looked like a tradesman's van but was actually a bit of a secret hot rod, a street sleeper. So, I needed a sporty engine.

Alfa Romeo Alfasuds were still quite common back then, but almost all of them were suffering from terminal rust. We got away lightly on the show with the 1982 Alfasud TI we featured; there was actually enough solid metal left to be able to weld in a patch ...

I found one in the free ads and relieved it of its powerplant. The 1500 Alfa Romeo engine was physically about the same size as a 1600 VW lump but produced around twice the horsepower and torque. It was crazy. Because the engine size was smaller, I actually saved money on my insurance. The Alfa engine's stud pattern and bell housing diameter were identical to the VW's; the clutch parts were interchangeable; even the coolant pipes lined up perfectly with the plumbing already in the van.

It all made for a really great, stock-looking conversion. When I went back to buy something else from the trader a few weeks later, he was flabbergasted to see that the van was on the road. He was even more dumbfounded to see that it now went like a rocket!

If twice the power was good, three or even four times would be even better. Following the success of the Alfa conversion, a Rover V8 was the obvious progression. I had already lowered the van by about four inches because I was testing a prototype lowering kit for Just Kampers, a Volkswagen parts and accessories company I've had a lot to do with over the years. The springs were deemed a little too low for retail but I loved them.

Lowering the van that much meant that the drive shafts now sat horizontally instead of at their usual incline from each wheel up to the gearbox. That, in turn, meant the CV joints were no longer running at an angle. Which meant that I could now move the engine and gearbox forward into the van by four inches without compromising the integrity of the CV joints' designed tolerances.

That was ideal, because when most people had tried a V8 conversion before, it had ended up hanging out the back of the van, which looked awful. With a few other modifications to pulleys and ancillaries it meant that I could have the Rover V8, with all of its extra power and torque, fully enclosed in the bodywork. I had built a proper street sleeper – it looked stock, albeit lowered, but packed a real punch.

While trying to shoehorn everything in, I found I needed an older style P5 distributor that was smaller than that on the SD1 engine I was using. I tried all of the local motor factors but with no joy. My last chance locally was a well-appointed little car parts shop in Yateley, situated right next to the HQ of the Monster Raving Loony Party, which was a pub, *The Dog and Partridge*.

Unfortunately, they couldn't help but they suggested I pop up the road to talk to a local chap who was building a Rover V8 drag car in his garage. I knocked on the gate of his drive and, as luck would have it, he was indeed working on a Ferrari-yellow TR7 with huge flared wheel arches and massive tyres. This guy was serious.

After a bit of a chat and a rummage through his stash of parts, he produced exactly what I needed. We shook on a deal and I rushed home to fit my vital part. It didn't take long to get the engine up and running again so I took it out for a drive. The van now had terrible acceleration, nothing like as good as before. Something really wasn't right.

I pulled out the dizzy and took it straight back to the bloke who had clearly stitched me up. Turned out, it was one of his racing parts and he had welded up the bob weights which advanced the timing as the engine spins faster. He apologised, found me a replacement and offered to pop over to give me a hand setting up the timing with his fancy strobe light. The new part worked a treat and from that moment on Paul Brackley became a great mate, eventually joining me on *Wheeler Dealers*, and making on-screen appearances as the series' technical adviser.

The V8 had plenty of oomph but it could be a thirsty beast. The van was originally a diesel so had both a fuel line and a return. Rather than block one of them up, I connected a small electric fuel pump and filter to both lines. To keep the dashboard looking stock I fitted the fuel pump switches and a gauge into the pop-up ash tray so they could be hidden. The engine was quite happy to run on one fuel pump for most of the time but if I really wanted to accelerate, it would suffer from fuel starvation so, I would simply switch on the second pump. It was great for fuel economy.

I zoomed around in that van for a while and then thought that the next stage would be to put a bigger, American V8, further inside the vehicle, mid-mounted. I saw an advertisement for a T25 V8 conversion 'unfinished project' for sale in Wales, so I hooked up my trailer and set off to see it.

It was, shall we say, an *interesting* conversion. The guy had turned the sliding door into a kind of gull-wing door by hinging its top edge along the edge of the roof. Two pairs of gas-struts supported the weight of the door when open and when it rained, water just poured into the van. It looked horrible and it was really impractical. He had also fitted a Jaguar back axle to take the power from the V8, but there was a lot of work to be done to make sure it wouldn't tear itself from its mountings once it was up and running. Definitely something to walk away from.

So, I bought it. Well, I had driven all that way and the guy was so desperate to get rid of it. When I got it home, my mates took one look at it and laughed themselves silly. 'What are you going to do with that? The door is horrible, you'll have to cut it out.' Cut it out? That sounded like a great idea, I could make a shorty van.

It took a lot of careful measurement, very careful cutting and a massive amount of reinforcement and welding but I did cut out the entire sliding door section of the van and turned the Transporter into a very cute looking 'Trans-Shorter'. It looked like a T25 that had been squeezed in a vice. The crash protection on the front of a T25 is very substantial and, instead of the planned V8, I ended chucking in a spare, not very heavy, 1200cc engine I had lying around.

On its first outing I drove down the hill near Mum's house and had to brake suddenly to avoid some traffic. The back wheels lifted off the ground and suddenly I was mid-endo. The back end started to swing out into the on-coming traffic and before I knew I was heading directly at the pavement. Any more braking would only make things worse. Thankfully the wheels hit the curb and I missed someone's garden wall by millimetres.

I calmly reversed back into the road and drove straight home to add some ballast. I ended up selling it to a VW specialist in

Ripley and he re-painted it and sold it on to a drummer, as, apparently, it was the perfect size for carting his drum kit around, and great for parking outside gigs.

In the early episodes of the show we had a very tight budget with which to buy and refurbish a vehicle. This started at £1,000 per car and increased with each series. We had been keen to do a camper van for a while but had to wait until the third series before our improved budget of £3,000 allowed us to buy a standard T25 van and then convert it to a camper.

It was the least desirable diesel version and in the worst colour choice of 'hearing-aid beige', but that is why it was affordable. We finished off the conversion by transforming its appearance with a vinyl wrap using a fun blue bubble motif. I ended up buying it and had a bit of fun with it over the summer, but one of our cameramen really wanted it. His wife was an artist and needed to move big canvases around, so I sold it to him.

Many months later, Paul Brackley's son, Sam, was working as a runner for the show and, while out on some errand, he had been stuck in traffic on the M3 for hours. When he eventually got to the cause of the delay he saw our bubble-wrap van in the layby looking very forlorn and rather scorched. The engine had caught fire, causing our bubble wrap to bubble up and burn too.

I am sure it was nothing to do with the new glow plugs I put in! If anyone is looking for a project, apparently the van was last seen in a scrap yard somewhere in Gosport.

* * *

The split-screen van, whatever the iteration, has always been *the* one to have. Whether it's the hippy-chic, the surfer-cool, or simply its rose-tinted history, their desirability means they can now command a huge premium. Right now in California, you can pay $150,000 for an electrified example. But many moons ago, probably 25 years or so, I found myself regularly driving past a ramshackle farm who seemed to be using an old VW as a kind of shed.

One day my curiosity could stand it no longer and I drove down the lane, tentatively knocked on the door of the mobile home, and asked if I could take a closer look. It was indeed a VW, in fact a split-screen microbus, but they had been using it for some time as a chicken shed. The mess was disgusting. Somehow, after a bit of chat, I parted with £125 and the farmer agreed to deliver the vehicle to Mum's house.

A few days later, it arrived on the back of a truck and was deposited, by hydraulic arm, onto Mum's drive. As you would expect, she was really impressed. Thankfully, the farmer had already removed the chickens and much of their mess but it was in a terrible state. The thing hadn't rolled, let alone driven, for years. Everything was seized solid; the wheels were immovable the suspension didn't move, the doors opened but the floor and much of the structural metal was missing.

I figured I would start with a look at the engine. I changed the spark plugs, connected up the fuel line to a Jerry-can and added a fresh battery. I turned over the engine a few times until the oil pressure light went out and while still sitting behind the engine, to operate the throttle by hand, I got someone to turn it over. On only the second rotation, the engine spun into life. Astounding! After all these years.

I was instantly enveloped in decades of soot and rust from the exhaust system, leaving an inverted shadow of me on the tarmac. I was absolutely covered from head to toe.

After a few more days struggling to free off the brakes, steering and suspension, a local firm offered me good money for the front of the van, so they could save another. Someone else wanted the roof to save a 21-window Samba bus. In the end I relented. My first splitty was not to make it onto the road, but breaking it up for parts and panels did rescue a few other vans and gave me a few quid to spend on my next 'investment'. Looking through websites now, you couldn't buy a van in even worse condition for less than £10,000. How times have changed.

While at uni, I shared a flat with Dave and Fred (also from my Engineering Product Design course), on Rainbow Street in Camberwell Green, London. The flat was above a chemist and over the road from a Chinese takeaway and a bakery – we were sorted. I would park my roofless beach buggy outside for most of the week and cycle to and from South Bank University, darting through the constant traffic.

My room was the biggest and became the lounge, so obviously we needed a TV and games console. At some point I had come into possession of another split-screen van front; I think it had been used as part of a display in a parts shop. I added a box section frame and a shelf, to turn it into a TV cabinet. We put a

TV in one side of the split screen and a computer monitor in the other so we could play games and watch TV at the same time.

For extra comfort, we sat in a pair of old leather front seats from a Jaguar, the epitome of student luxury.

When it comes to the epitome of automotive luxury, you might plump for a Rolls Royce or Maybach. But when you think about it, sure they have heated, massaging, reclining seats, well-appointed bars, plenty of legroom and sublimely deep carpets, but if you want a king-size bed, washer/dryer, full kitchen, bathroom and shower then you want an RV. For maximum space, features and downright gaudy opulence you need an American RV, built on a coach chassis.

Imogen and I met a chap who ran an RV business at an event once. A while later, we were passing his place so we popped in, just for a look. There were loads of different styles and sizes, but the huge American coach-based behemoths were something else; a palace on wheels. Nobody needs an RV that big, but somehow we left there the proud owners of a 36-foot-long mobile house with a couple of pop-outs that made it even bigger once parked up.

Technically, they are too wide for British roads, but they get special dispensation and, just to add to the obscenity of our new camper's road hogging, we hitched on a huge covered car trailer too. It made for very convenient and comfortable living. Camping is great fun, but when the English summer delivers cold rain, again, it is lovely to re-charge in a warm and dry living room with a large, well-stocked fridge!

My first wheels. I'd absolutely refuse to lie down in my pram. I was on my belly, head up into the wind, taking everything in...

Once self-propelled, I'd tri-cycle round to my neighbour, Maria's for lunch.

Mum & Dad brought home a sister, but at least they got me my first big-boy wheels.

ting up my irst garage, t may have een Clare's, but as a big brother, surely it was my job to show her how to play with it properly.

Dad's lapel p
from when h
worked on th
Ariel satelli
project

With Mum and my favourite red ball
the beach in Cornwall. I once manag
to 'lose' that ball out of the car
window, right at the top of a really
long hill on the A303. I made such a
fuss that Dad had to park up on the
hard shoulder and walk all the way
back down the hill to get it for me.

In my school uniform from South
Farnborough Prep school, before it
turned into Rushmoor Independent.

Family breakfast at
my grandparents' in
Cornwall, Dad had,
yet again, foiled
an attempt at the
cuckoo clock...

In the CDT workshop at King Edward's, with the masterful Mr Bullock and Mr Murfitt, (who used to chuck me out of physics).

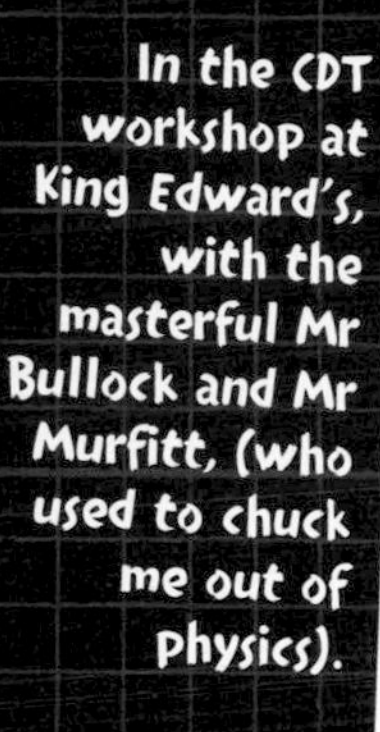

A smattering of my 'extracurricular projects': what's left of one of the 'prowling house-master' warning systems and the 'A' key for the chem lab made from a photocopy using a key blank and a needle file. It worked a treat!

Testing my land yacht on the beach, using Auntie Fran as an anchor, while I'm fixing the rigging. Just look at that comfy seat, hand stitched by Mum.

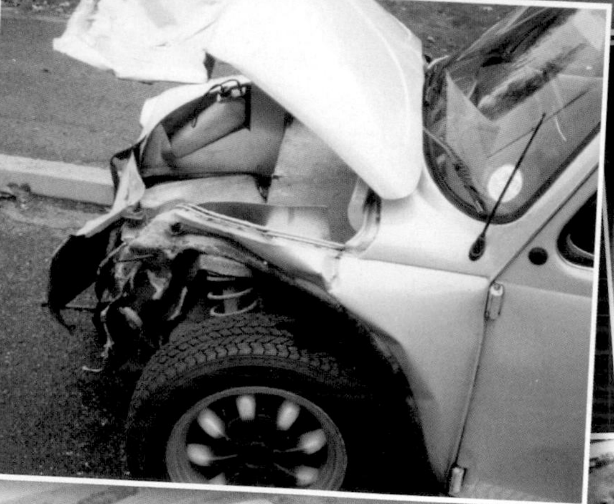

My 1303 Texas Yellow Bug – three ways: pristine pride & joy, crumpled beyond repair and reborn as an ugly beach buggy.

'Eddchin
perfecte
rusty met
fused to yo
window c
be cool if y
mask u
design fi

With Clare,
Auntie Fran
and Yanna th
dog. Who ne
a windscreen
when you lo
this good in
flying goggl

ne mighty Leyland tlantean, trying to fathom the inner workings the huge l engine.

My second land yacht, the RAWV, and its lanky creator. This one was amphibious and got me through my degree.

My Fugitive sand rail was just a pile of bits but Mum could see that it had potential. She always did have absolute faith in my vision...

Kicking up a sandstorm in my Alfa Romeo boxer powered Fugitive. I ripped off the exhaust on a tree stump moments after this shot!

There weren't enough VDubs on Mum's drive, so I thought I'd add to the collection. With the split-screen chicken shed...

Lovely pub for a spot of lunch. Spot the excellently parked RV.

Dave and I setting off for Run To The Sun in the 'teddy bear car', The Furry Ugg.

Doing an 'endo' in the Transhorter.

The Dutch Army camper van, with my custom made roof rack.

Mum's first garage and my first workshop. I think I've utilised the space quite well...

Don't know why the neighbours made such a fuss...

Love at fi
sight, my
scrapyard
firetruck i
the Califor
High Deser
It's still or
the list...

We once took the RV to Ireland for Imogen's Auntie Rosie's birthday party. Imogen's mum and dad joined us, and we headed for the Pembroke to Rosslare ferry – and managed to miss it. No drama, we were there to relax and we could always get the next ferry.

It gave us a few hours to kill in South Wales and we decided to go for a drive. We knew we were limited in where we could go on Pembrokeshire's tiny country roads, so when it was time to stop for lunch, we phoned ahead to a pub to make sure they could receive us.

I explained that we were in an RV the size of a bus and asked if they had a coach park. They said that yes, no problem, they had two car parks and plenty of room for a bus. Excellent, we were hungry and ready for a nice pub lunch. They warned us that the road was a bit narrow but 'don't worry, just keep coming and we're right at the end of the road. You'll see the car parks either side of the road, just before the pub.'

Well, they really were right at the end of that narrow, winding road. We had inches to spare on either side, all the way down, but the RV had good mirrors and cameras, so we crawled on. The pub was split either side of a slipway; the Atlantic Ocean was literally lapping at the tarmac. But there was no coach park. There were three car spaces on our right and perhaps a dozen on our left. All of which were full.

When I had said bus, meaning coach, she had heard bus meaning camper. Bugger! There was only one way back out, I was going to have to reverse most of the way back up the tiny winding road and find a space to attempt an 'Austin Powers' style gazillion point turn. But not on an empty stomach.

I edged forward as far down the slipway as I dared, allowing any parked cars to leave, but wedging the RV further between the two pub buildings with the mirrors inches from the walls and parked up. It was time for lunch.

We had a lovely meal by the sea and started reversing our way back to Pembroke Dock, dodging parked cars, ancient walls, bushes and over hanging trees. We did take a considerable amount of the greenery with us but only squashed one of the roof-mounted airhorns.

By now, many of the other visitors had vacated a third car park further up the hill, which made it rather easier to do my many-point turn. After getting a few more cars moved and a lot more shuffling, we made it. Success, and great practice for the even smaller roads of West Cork.

We sold the RV when we were relocated to California; it was one of the many things that had to go. I was sad to part with it but it went to a good home and I was very pleased to see it again at a race meet at Spa circuit a few years later. Adventures in a camper, however big or small, are something that will always appeal to me, so I am sure it won't be long before we set off on another. In fact, I will guarantee it!

4.

BUG-GERING ABOUT

It has been said, on more than one occasion, that throughout my life I have been mostly just buggering about. It is difficult for me to argue with that accusation. Joy and happiness come from not taking things too seriously.

Back in the days when I could hide in a crowd in plain sight, the way a tree can hide in a garden full of gnomes, I used to go to a lot of car shows with my mates just to see what everyone was up to. Run to the Sun was one of my favourites. It was a 'cruise' of hundreds of vehicles from London to Newquay, followed by a weekend of messing about in cars, camping and a 'Show & Shine' event.

One year back in the day, perusing a post-event article in *Custom Car* magazine, Steven and I noticed we were in the background of a photo – well the back of our heads and one of our shoes, at least. This taste of infamy spawned an idea; why not do something worthy of an actual photo in next year's report?

As the name suggests, a Show & Shine is usually populated by loads of beautifully restored and customised cars, and the winner was sure to be featured in the magazine. However, that would need a lot of effort and usually a lot of money. So, I figured the best way to stand out from the crowd was to step away from the crowd; I had so many crappy cars lying around perhaps taking that idea to extremis could work.

My embryonic plan was to create a kind of Flintstones' car by removing the windows and roof of one of my wrecks and then perhaps paint the bodywork to look like it had been hewn from granite. Thinking about it now I guess it was an early kind of a Rat Rod, positively Neolithic. Dave, Steven and a few other mates came around to start work.

I marked out the lines I needed to cut to remove the roof, leaving a little 'racing screen' to protect us from the elements. At some point during the proceedings it occurred to me that costumes would be a great addition, so we dropped everything and rushed to the nearest haberdashery to buy some animal print faux fur for our caveman tabards.

There was a lot to choose from, though leopard print seemed the most fun, but then something else caught my eye. The was a simple brown fur called 'Wolf'. That was it! I suddenly knew we needed to cover the caveman bug in the 'Wolf' fur, it would be perfect. After a rough calculation, I bought the entire bolt of fabric and also an offcut of faux ermine to cover the screen.

Back at the workshop, my happy helpers were getting a little too happy as the alcohol flowed, so the welding of the doors to their frames, for strength, and capping off of the cut metal edges, was left up to me. Over the following week or so, I fixed everything needed to get it through an MOT and Mum sewed up the costumes.

To add the fur, Dave and I got busy cutting pieces of the Wolf to roughly the correct size and then coated the car and the fur with contact adhesive. It was a simple process but it took ages. And then it started to rain. We hurriedly built a ramshackle tent by parking a van on the drive and one on Mum's front lawn and then stretched a tarpaulin between the two. We carried on sticking into the night. We caught a few hours' sleep and by the morning we were ready to set off in our caveman costumes and flying goggles, and thanks to the glue and the fur we were now sporting rather furry palms.

We set off to join the cruise down to Newquay and while waiting on a motorway bridge over the M4 near Reading we turned on the radio. The radio had come with the car, probably from the factory, and even though the aerial was connected we could only get one station – a pop music station from France, in French.

To our delight, the first song to play was Duran Duran's 'Hungry Like the Wolf', surely a good omen; this was going to be a great adventure.

We had a brilliant weekend with so many people coming up to chat wherever we went. After a lazy day of aimlessly driving around Newquay, we stopped off in the high street to order a Chinese takeaway. A local policeman came over to inform us that our antics had been followed all day on their CCTV and, even though it had given them a lot of smiles, they couldn't ignore the fact that we were waiting on double yellow lines. We needed to move on.

We moved our 'dinner party' to a car park around the corner and enjoyed our oriental feast in the evening sun. While we were eating, a car pulled up behind us and a family got out and explained they had been following us around the town much of the afternoon and asked sweetly if it would be OK for their young daughter to say goodnight to the 'teddy bear car'; apparently, she wouldn't go to bed until she had. Of course!

Holding her mother's hand, the little girl coyly sidled up to the car and then excitedly flung her little arms around the bonnet, giving the car a huge bedtime hug. It was so heart-warming to see, I actually started to well up. In the eyes of this little girl, my furry Beetle was a living teddy bear. Suddenly I felt like Walt Disney must have done, capturing the imaginations of children everywhere by breathing life into an inanimate object, giving it some kind of soul. That was a really special moment.

The next day at the Show & Shine, there were the usual rows of gorgeous, gleaming cars on display. Our flurry of fur was never going to win an actual prize in that company, but we did bag ourselves the photo in the magazine we had hoped for. We also caught the attention of a team from *Top Gear* magazine who organised an impromptu photo shoot on the sand of nearby

Porth Beach. The caveman inspired scene featured me dragging a friend, Frankie, by her hair, alongside the car.

It wasn't only children and journos who were fascinated by the Furry Ugg. That summer, we had friends over from the US and I gave them a tour of the general area, including a cruise past Windsor Castle, to let them have a taste of England. Of course, I took them in the Furry Ugg. They thought the car was quite amusing but were over the moon when we were pulled over by two English police cars. I was a little concerned when four female officers jumped out rather enthusiastically, but all they wanted was to question how stroke-able was the Wolf fur.

Despite the Furry Ugg's silly beginnings, I felt it somehow made the world a better place, as it raised people's spirits. Then again, high spirits or showing off can often lead to making an arse of yourself, a concept I'm not entirely unfamiliar with.

* * *

By Series 7 of *Wheeler Dealers*, the crew and I had spent many years together, stuck in the close confines of the studio. To stop ourselves from going completely stir-crazy we joked and arsed about, a lot. To mix things up for one of the episodes, the producers had decided it felt natural for me to do a mini-test drive, mid-show, to sample the new gearbox I had replaced on the brilliant, orange, three-wheeled wedge that is the Bond Bug.

The crew set up the camera just before a 90-degree, left-hand bend at the end of a straight country road. I hid around the corner in the distance waiting for 'Action!' over the walkie-talkie. As usual,

I was wired for sound with a radio-microphone so the camera and sound man could hear my lines as I drove through the shot.

When Tom Karen, the genius designer of the Reliant Scimitar and the Raleigh Chopper, designed the Bond Bug 40 years before, he'd had no idea that it would become the basis for the Landspeeder in the first *Star Wars* movie, nor that it would ever need to accommodate a Wookie sized, 6-foot-7 mechanic. I really struggled to fit into the car. The pedals were so close together that I had to drive in socks, and I had to hook my leg over, rather than under, the bar that supported the steering so I was able to operate the clutch.

'Action!' I got up to speed and whizzed around the right hand bend into shot. My weight, enough to counter the roll of the body, keeping the Bug's three wheels safely on the ground. I could see Bear, the sound man, standing behind the camera with his boom, grinning as usual. We already had a number of shots 'in the can' and driving the Bond Bug was so hilarious I was in a frisky mood. We were both giggling, and by the time I was through the frame of the shot I could see him clearly egging me on; he knew what I was thinking.

I jabbed the steering momentarily to the left and, thanks to the imbalance of my weight, the Bug lifted its passenger wheel into the air. I was now careering towards the end of the road on two wheels, laughing all the way.

The sharp left-hand bend was fast approaching so I tried to slow the car. Clearly three brakes were better than two; I was still too fast. Nudging the steering to the right might have lowered the cocked wheel back to the ground but then I would have been on a collision course with the hedgerow. I was gradually losing a bit of speed so I went for broke, hoping I could edge around the tight corner, on two wheels, with a wide line.

Strangely, that didn't work out, I was already leaning too far over. A few more degrees to the left on the steering wheel sent me past the point of no return and the Bond Bug casually tipped over onto its side, scratching and scraping along the tarmac to an undignified stop in the grass verge next to Bear, who by now was wetting himself with laughter.

The boys helped me tip the car back onto its wheels and we finished the shots, careful not to film the damage. Once back at the workshop, we got on with the next bit of filming and Paul Brackley saved my blushes with a dab or two of filler and some fresh orange paint.

The Bond Bug is a really fun, quirky car, but parts are increasingly difficult to get hold of. Because only 2,270 were made over its four-year run it is unlikely anyone will go to the trouble of tooling up for replacement parts. Conversely, some 21 million VW Bugs were made over their 65-year run and most of the parts are interchangeable, even across so many years, so they are easy to get spares for and so are a particularly great first car to learn on.

* * *

After my first Beetle, I had certainly caught the Bug bug, but such hobbies need funding. To finance my growing fascination with all things automotive, I took on a variety of part-time jobs.

I'd had a great time working at Pizza Hut but it became clear that there were other ways that I could spend time learning about cars while making enough to pay for petrol and my slowly expanding collection of tools which, mindful of how useful it had been to get the Texas Yellow Beetle past its MOT, soon included welding gear.

When I went to buy my first welder, I asked the guy if he could give me some sort of training on how to use it. He adjusted the wire feed and current setting on my new bit of kit, stuck a piece of scrap metal in front of me and said, 'Write your name on that.' After a couple of goes I had produced some squiggles which just about looked like my name, if you squinted a lot.

'There you go, see?' he proclaimed. 'There's nothing to it, you'll be fine.'

Word soon got around that I had a welder, and Will and Bob, the blokes up the road who were building the Westfield kit car, came down to give it the once-over – and help further my training by showing me how they would fix their exhaust using my welder. They wanted to close up a tear in the exhaust, but maybe because the settings were off on the welder, or maybe because they were too busy bantering and taking the piss out of me to pay full attention, every time they tried, they just blew a bigger hole in the exhaust.

So, to spare them any further embarrassment, I sidled up, fiddled with the settings on the welder and had a go myself. The pressure was now on to not make an arse of myself and I put every ounce of my focus and concentration into filling that hole up as neatly and prettily as possible. I presented their beautifully repaired exhaust back to them with all the humility that the situation warranted (none at all). I had now earned my place as an equal in their banter. Spirit of Ignorance indeed.

After that, I had one more practice, repairing the battery tray on Steven's Beetle, then felt ready to tackle my next big project. My plan was that my yellow Beetle, still sitting forlornly on my Mum's driveway with its front caved in, should be given a new lease of life. I was going to build a beach buggy.

The front of my Bug was a write-off, but the rear end had survived unscathed, which was good news. On a front-engined car, the radiator and cooling system, alternator, battery, maybe even the engine might all have been junked. However, under the bonnet of a Beetle lie just the spare wheel and the fuel tank, and being air-cooled, there is no radiator. It meant there was a lot of the car that I could re-use once the accident insurance was sorted out.

When I got the insurance money, I bought a 1302 Beetle shell with a very rare, factory-fitted sunroof. I preferred its more minimalist styling to that of my old 1303. It had the more typical flat windscreen and metal dashboard of earlier Beetles but still had the independent rear suspension (IRS) and Macpherson struts at the front. My plan was to swap all the working parts of my old yellow Bug into this shell to get one working car.

However, as I dismantled the new shell, I started finding more and more very well-disguised rust. The body shell and floorpan, rather than being bolted together with a gasket in-between as they should be, had been welded together in an effort to disguise the rot in the heater channels. Stripping out the rot was turning out to be a huge job so the 1302 remained dangling from my Mum's garage roof for quite a while. A job for another day.

I was getting impatient, so I prised the mangled body from the slightly stunted floorpan of my 1303 in readiness for a new plan. I then bought a slightly older, 1200 Beetle, because it had the original Beetle beam front suspension, as opposed to the Macpherson struts. Macpherson struts were great for road driving, but the beam suspension is more robust and far better for off-road.

So, I built my first beach buggy using the rear half of my 1303 floorpan with its IRS, gearbox and proven 1600cc engine, and the front half of the 1200 floorpan, with its rugged beam axle front suspension. It was the ideal combination.

The 1303's IRS is far better than the 1200's swing axle rear suspension for stability and transferring power to the ground. When you see older Beetles with the swing axle suspension taking corners at speed on their thin little wheels, the terrible positive camber of the suspension kind of tucks the wheels under the body, and that can easily cause the car to skid or even roll in extreme conditions.

For the bodywork, I went to see a guy called James Hale, who had a company called GT Mouldings. They produced a fibreglass beach buggy body that was probably one of the ugliest beach buggies known to man. It was a Baja GT. It had headlights right at the front in a sort of nose cone. I went for the ugly bug because it looked more robust, needed less chassis support and it came with a roof that I thought might come in handy. I had some SAAB seats which I narrowed down a bit to make them fit. They were nice and comfortable but a bit rubbish in the rain. I guess that, if you are going to drive around in the UK with no roof, no windscreen and no windows, you should have seats that let water drain out of them rather than soaking it up like a sponge.

For ages, I never bothered putting a windscreen on my buggy. It was my daily driver but I used to hurtle around in it with just flying goggles and the wind in my hair, along with various bugs, bits of tarmac and a lot of road dust. I still have one pair of goggles with a broken lens where a stag beetle hit me in the eye. I was very glad of the goggles that day.

It was noisy, too. I still have tinnitus from the sound of the wind rushing past at speed. Well, that, and angle grinding without earmuffs, but the worst was the rain. The only way to deal with a heavy shower of rain was to try to suck as much of your face as possible into your mouth to stop the rain from stinging. You have to make the choice: do I go faster to get home quicker, or do I go slower so that it doesn't hurt so much? Generally, I would choose the faster option. The things that get chucked up off the road into your face are to be expected and most of the big stuff, traffic cones and the like, you can generally avoid.

I remember driving it down to Cornwall one summer and for much of the length of the A30 I had to endure the most horrendous torrential rain storm. I could barely see the road ahead, so had to rely on the lights of the vehicles in front. I was completely engulfed in a squall of water for nearly the entire journey. Somehow the engine managed to run. Despite breathing in as much water as air, it just about stuttered along, so I didn't dare stop, just in case it wouldn't start again.

In the sunshine of the following day, I checked the condition of the oil only to find it was overflowing with a milky beige emulsion, just as if it was water-cooled and had a blown gasket. A quick oil change sorted the engine but it took a little longer to fit the screen, even though I now had plenty of motivation.

To support the screen and roof, I built a roll cage out of 50mm CDS (cold drawn seamless) tubing. I didn't have a pipe bender so ended up using cast gas pipe corners – not ideal by any means, but it was very strong and gave the buggy's styling new purpose. The cage also held the car together really well, making it far more rigid, improving the handling no end.

* * *

In those early years of experimenting with cars, I developed my instinct for how things worked and might go together by simply trying stuff out. I would ask experts for their advice and friends for their help, because dipping into other people's experience saves a lot of time and really helps to get a different perspective when it comes to solving problems.

I am often asked how I know so much about so many different kinds of cars. The answer is simply that it's an accumulation of knowledge over time; I continue to learn with each new project and rely on being able to pick other people's brains to speed up my learning.

Right at the start I gleaned a lot of information from magazines, books and my Saturday job at Just Kampers. There were many companies and shops catering for the VW scene at the time, so plenty of advice, parts, accessories and kits were always available, but actually seeing things from the other side of the counter was very useful and gave me even more access to information, cheaper parts, and of course, the opportunity to collect more unfinished projects.

I didn't make much money from my Saturday job – really just enough to buy my lunch with the guys on Saturday and keep me going in petrol for the rest of the week. However, I learned a lot, and when people realised that I was into messing about with old wrecks, I was offered all sorts of cars, often for free.

I saved up and bought a trailer for about £1,200 – a lot of money for me back then – and at this point my acquiring of dodgy old motors got out of control. I could now turn up outside someone's house and pick up all sorts of rubbish that I could drag home. I didn't always drag it to my own home. One of my friends, Malcolm, lived in a house in Farnborough with a very big garden, and his dad, Arthur, our family doctor, let me keep a few cars there.

Malcolm was one of the youngest people ever to build a replica GTD40 from a kit. He did it in his super tidy, lovely warm garage with lots of shiny tools, while I was just as happy out in the yard surrounded by my rubbish doing a bit of welding on some mate's car underneath a tarpaulin in the pouring rain (when it comes to health and safety, by the way, MIG-welding in the rain is *not* recommended!).

Having the trailer also meant that I could move other people's cars around – a car that needed to go to the paint shop, one that had to go in for engine surgery, or simply a motor that had to go from the driveway to a lock-up because its presence was no longer tolerated in the front garden. Some of this was for payment, some of it was trading favours, but all of it spread the word that I was a guy with a trailer who would happily sort out vehicles that weren't going anywhere under their own steam.

Sometimes that meant simply weighing them in for scrap. I'd also take any scrapped vehicles from Just Kampers and, at a time when the value of scrap metal was shooting up, I earned a few quid on my visits to the recyclers up the road in Blackbushe.

Those guys in the scrap yard had skills, and not just with their alchemy of turning any old iron into gold. I had cleared someone's drive of an old Rover SD1 – particularly handy as I needed a V8 for yet another VW van project I had embarked on. I quickly removed the engine on the road in front of my mum's house and in the process had cut a few useful parts off the car with my oxy acetylene 'gas axe'.

There were a few small fires as underseal and sound-deadening caught alight, but a wet rag soon put them out. Having scavenged everything I wanted, I strapped the car down onto my trailer ready for a trip to the scrapyard first thing the next morning.

I got to Blackbushe early enough to beat the usual rush and rolled onto the weighbridge as usual. The gross weight of my rig was noted and I was waved on. I swung into the yard, removed the straps holding the Rover down and reversed the trailer towards the grab crane.

The operator opened the jaws of the huge grab and gently lifted the car off the trailer as if the crane's arm was an extension of his own body. He popped it into a clearing in the middle of the piles of scrap around him, deftly flicked the car over onto its back and then plucked out the fuel tank like he was nicking an olive from a salad.

The moment the jaws split the tank, fuel sprayed out in every direction. Stupidly, I had forgotten to drain the tank as I was supposed to. In a split second the cloud of fuel turned into a fireball, enveloping the Rover, the crane and much of the scrapyard around it. Turned out, that the burnt sound-deadening had not completely 'gone out' and had, in fact, been smouldering nicely all through the night; the perfect source of ignition for a Rover flambé.

The crane operator didn't hesitate. He dropped the fuel tank, swung the arm to the edge of the yard, picked up a ginormous silo of water and deftly splashed it over the entire yard, instantly extinguishing the flames. His fire-quenching abilities were super-human.

I rather inadequately waved him 'thanks', drove back out onto the weighbridge, and extremely sheepishly crept into the reception to receive my cash for my deposit of scrap.

Unbelievably they actually paid me, but they certainly had a few other words to say too.

* * *

Over the years, I did quite a bit of work for Mark at Just Kampers. For a while, he would trawl the junk yards of California for rust-free VW parts like van doors and front beam axles to resell back in the UK. Paul and I even went out a couple of times to help him out.

Mark had noted on a road map all of the locations of scrap yards and useful businesses they had found the previous years. We would start each day with a big American breakfast in a diner, checking *Recycler* magazine for interesting cars and parts for sale and then traipse off into the 'boonies' in a Rent-A-Wreck van with a U-Haul trailer.

We would line the floor of a shipping container with the front axles, cover them with sheets of plywood and then load in the doors and any vehicles we had found on the way, filling any spaces with boxes of new parts.

On one trip, we were in the desert, down near the Mexican border, when I was completely taken by a wonderful 1940s fire engine that was being slowly reclaimed by the sand. As is often purported to be the case by junk yard owners, 'it had driven in', but now extracting it from the dunes was going to need a lot of effort. I resolved to return one day to rescue it. It's still on the list ...

We met some real characters on those trips, from all walks of life, but perhaps the most baffling were in and around the junk yards of the high desert. The first chap was a helper in a yard and was curious where we were from: 'England', he pondered for a moment ... 'Did you drive over?' – 'Yeah sure, windows up for the big waves!'

When grabbing a take-out lunch in a local burger joint one day, the waitress caught our accent. 'You don't sound like you're from around here, where ya from?'

'England?'

She thought for a moment. 'You've picked up the language quickly.'

'What language do you think we speak in England?'

She thought some more. 'French, isn't it?'

Well, perhaps news had just not reached that tiny part of a very big desert; we haven't spoken French in England since Chaucer's time – about 600 years ago ...

Cruising the road less travelled was always rewarding though. I found a Karmann Convertible Beetle, a Karmann Ghia and a wonderful split-screen crew-cab pickup truck with a sporty 1835cc engine. We had the crew-cab transported to our rented storage yard by the docks and drove it straight into the container from the low-loader. For the Karmann Ghia I chanced my arm and bought an A-frame and towed it up Interstate 405.

The Rent-A-Wreck van didn't come with a tow ball but had a pretty substantial one-piece bumper with a step. On closer inspection, I noticed that under a bit of grip tape there was already a hole, so I bought a tow ball from Pep Boys, punched it through the tape and bolted the ball into place. The A-frame clamped around the beam axle on the Ghia and onto the ball – it worked a treat. That is, until smoke started billowing from the front wheel of the Ghia.

I pulled over, worked out that the brake calliper was seized on and found a knife to cut the flexible brake pipe. Having relieved the pressure, once the brake had cooled down a bit we got back on our way. I sold the Bug and the Ghia pretty quickly once they were back in the UK but I smoked around in the crew-cab for many months until some chaps twisted my arm and bought it to promote their new venture.

Funnily enough, we echoed those early desert trips on the show some 10 and 15 years later when we bought both a lovely Karmann Ghia and then later a brilliant all black split-screen Microbus.

* * *

In the second series of *Wheeler Dealers*, the final car that we refurbished – well, built really – was a Predator beach buggy. It just so happened that we had found a rotting Beetle that was far too far gone for us to feature on the show, but rather than write it off, I suggested that we convert it into a beach buggy.

It was our biggest project to date, and the producers saved it until the last car in the series. As was typical for all the cars we restored, what we were able to cover in the show was only a fraction of the actual work necessary to get the buggy finished. It was probably three to four weeks' work condensed into the usual 22 minutes of TV.

In the early days, once the camera crew went home for the day, I had to set to work, completing jobs and preparing the car for the next day's filming segment – it wasn't until series 3 that I convinced the producers to let me bring my friend Paul Brackley on board to give me a hand. So I had to work flat out for the last fortnight, to get the buggy finished and MOT-ed in time for the test drive on the wonderful sandy beach at West Wittering.

The new owner was going to drive it away after the sale but I still had to get the beach buggy to the coast on the morning of the shoot. For some reason, I didn't have a spare trailer or a vehicle with a tow bar available. I did have the old A-frame I had used for the Karmann Ghia, and a promotional car that I was tweaking for a customer had a tow bar. Problem solved. So, if anyone saw a flying saucer towing a beach buggy towards West Wittering sometime around 2004, now you know what was going on.

I had a couple of other beach buggies, and I even bought one that was allegedly used in a James Bond movie, *For Your Eyes Only*, to try to run down Roger Moore. But one of the buggies in which I had the most off-road fun was my Fugitive sand rail. A sand rail is a different kind of buggy that doesn't rely on using the Beetle floorpan. It has its own, very strong, lightweight tubular structure, like an exoskeleton.

I bought mine as a kit of disassembled parts, which included a few fibreglass body panels. The sand rail had been given a very dodgy paint job in its past so I needed to get it sand blasted. This seemed like a great time to build a very strong roof rack for my Dutch Army van, so I found some box section steel that fitted perfectly in the van's roof gutters and set about constructing a roof-rack fit for, well, almost anything.

It didn't take long to weld together and ended up looking a bit like an iron bridge. It was certainly strong enough. I was wondering what colour to paint my Fugitive so I took the opportunity to experiment on the roof-rack. I didn't like any of the standard colours in the Hammerite range so I mixed their red and blue together and created a rather fetching purple which looked great against the drab green of my van.

Once my structural roof-rack was complete and bolted on, I lifted the sand rail frame on top, lashed it down and set off to the sand blaster.

As I drove home, the heavens graciously sprinkled my freshly peened frame with a light drizzle. The blast media left the surface of my tubular frame with millions of tiny dents, brilliant as a key for the new paint but terrible, thanks to the huge surface area, for corrosion. It only took 15 minutes to get home but, in that time, the naked steel surface had already turned into lovely rusty orange.

I had already mixed up a new batch of my special purple paint so I dried the frame with a towel and got busy with the protective coat before I needed a second trip to the blasters. A few weeks later, the Fugitive was ready for the road, but it was going off-road that I was most interested in ...

The Fugitive was very capable off the beaten track, and not too far from home were acres of Army training grounds, 'The Ranges', where I could hurtle around in the scrub and woodland without disturbing anyone. Well, except perhaps the Ministry of Defence.

I once cut through some land near Mytchett, and as I blasted around a corner, I was met by a platoon of armed squaddies on a training exercise. We were equally surprised to see each other, but they immediately set their gun sights on me and I spun the car round and got out of there in a rooster tail of sand. I could hear cracks of gunfire behind me, but I am pretty sure they were blanks. I didn't use that short cut again; there was, after all, plenty of other MoD land to enjoy ...

5.

SOFA, SO GOOD: THE CASUAL LOFA

While I was at uni, I began to accumulate a collection of valuable classic vehicles. Or, as my mates liked to call them, 'rusty old shitters'. At one point, I had around 35 different car and van projects on the go – not because I planned it that way, but because cars simply kept presenting themselves to me. Let's face it, nearly every car has something good about it that you want to take a look at, take to pieces to see how it works, or maybe re-use in another car.

Admittedly, the fascinating bits of some of the cars that I had weren't always easy for others to appreciate. The neighbours certainly weren't the biggest fans of my growing accumulation of clutter, although Mum was reasonably tolerant of it. She and Dad had had their two Minis on the driveway for years, while they worked on them, so she knew the way these things work.

'While I was at uni' also covers a fair period because, as already mentioned, I was there a little longer than I originally planned. This was mainly due to the fact that I changed courses after the first year and then, during my final year, I was very distracted with Other Stuff, mostly all my car projects and the jobs I had to fund them, so my lecturers suggested that I defer for a year.

My problem has always been that I am too easily distracted. There is just so much interesting and fascinating stuff in the world to explore. At uni, I usually found the projects other people were working on far more interesting than finishing my own. When they hit a brick wall trying to solve an engineering problem, I just couldn't resist getting involved.

After spending lots of time on my own projects and challenges, looking at new things seemed far more interesting than getting my own stuff down on paper. Completing stuff is not my strong suit; solving problems is. So, in short, for a number of reasons, getting my degree took longer than expected, but in truth, it was always just the backdrop to all the other projects I got up to.

The first car after my beloved yellow Beetle was the Fiat 132. It came from one of my friends at uni. This guy had one of the most awesome jobs imaginable for a student. He worked for a company that delivered supercars to superstars. If someone from Frankie Goes to Hollywood or Pink Floyd ordered an exotic set of wheels, he would drive it to their house and park it on their driveway.

When this mate heard about the Beetle/Rover disaster, he told me that he had a Fiat 132 sitting outside his house that was a non-runner. He gave me the keys and the V5 and said it was all mine as long as I could drag it away. I persuaded Mum to give me a lift to his house a few days later, taking with me a fully charged battery and a can of petrol. That was pretty much all it needed. After a bit of fettling, the engine fired up and I was the proud owner of a Fiat 132 saloon. It was an automatic, but still quite sporty, with the two-litre Mirafiori engine.

But if you are thinking that I smugly cruised home in a nifty little sexy Italian number, think again. Inside, the car was seat-squelchingly damp and streaming with condensation. The windscreen was constantly fogged up because the fan blew about as hard as the Pope's last breath. The headlights were hopeless and the windscreen wipers smeared, rather than cleared the view ahead.

So, I had to follow Mum home in a light drizzle – it all felt more Lake District than Lake Como – with my head out the window to keep her tail lights in sight. We got there eventually and I dried out the car, fixed the problems and got it through an MOT. I was mobile again ... well, until the Fiat did a 'General Lee' into oblivion. With most accidents, there is usually some element of losing control, even if it means someone else's actions that you were not in control of.

As a teenager at school, during our A-level years, the staff organised bar evenings, presumably to help teach us how to be sensible with alcohol. I have never liked beer and didn't yet appreciate wine so I swapped my two raffle tickets for two cans of Strongbow cider. Someone suggested we try 'shot-gunning' our drinks – so, obviously, we did.

Although I didn't smoke, I joined some friends on a short trek to the woods so they could partake in their illicit pursuit. By this time the alcohol had kicked in and I noticed I was saying stuff and doing things, like climbing and jumping out of trees, that I wouldn't normally do. In short, acting like a complete arse.

This lack of self-control bothered me, so I just decided to try to stop drinking, first for a week, then a month, then a year, to see what would happen. In the end I didn't drink again until my thirtieth birthday, when my friends bought me pretty much one of everything on the bar menu.

The upshot of this 'life choice' was that when my mates wanted to go out, they always had a designated driver. I was very happy with this arrangement and over the years it meant I got to drive a lot of interesting vehicles and usually to where I wanted to go.

So, one Friday evening in my late teens, five of us were packed into the Fiat 132, as we headed into Godalming. One of my mates, Graham, was sporting a broken leg in a plaster cast, and we were on our way to his house.

We were driving through a hilly area, somewhere near the Winkworth arboretum, on a typical country road flanked by verges and hedgerows. The road had a really long, right-hand bend immediately followed by a tight left-hander, suddenly dropping away down a steep hill.

My Fiat was a fairly cool car and quite quick. We once took it up to 107mph on the A3 in second gear – mainly because I had neglected to put the gear lever in 'drive' and was trying to make it change up. The 2.0L twin cam was a strong little engine.

That evening outside Godalming, we were all chatting and having a laugh when we rounded that particular tight left-hand bend – 'The End of The World', as the locals called it. I reckon we were probably doing about 50mph. Not technically speeding, but definitely going too fast for that section of road. The car took the bend and then, suddenly, we became weightless.

We were well clear of the tarmac and sailing through fresh air. It all happened in a few seconds, but it seemed to last an eternity. The car was airborne and all of the chat stopped. The engine was thrumming away but there was no longer any road noise. It had all gone quiet. Even the stereo seemed to have turned itself down. I remember thinking, 'Wow! I can see the top of the hedgerow. That's not right, is it?'

As the wheels were no longer touching the ground, the steering had little influence on our direction so I resorted to sheer force of will to get the car to waft away from the verge and back onto the road. We landed with a huge bang. There was crunching, scraping and screeching, as the force of our impact steered the car back towards the verge.

I wrestled with the steering wheel, which was still not really having much effect. We shot up the bank and bounced off the shrubbery back down onto the road, where the car ground to a halt, and there was silence again. We took a moment to understand what had happened. Someone gingerly asked 'Are we dead?'

'Is everyone OK?'

'Yeah.'

'Graham, are you OK?'

'No. I've still got a broken leg.'

We all got out of the car and got Graham out with his plaster cast. The doors were still opening OK, but there was broken glass around. The headlights were smashed. Apart from that, the car didn't look too bad.

Except, the front wheels were toeing in massively. When the car had returned to earth, the weight of the engine had pinned the steering's drag link between the oil sump and the tarmac. As it was dragged further under the car it bent into a U shape, forcing both wheels to point into the centreline of the car. We could see liquid coming out from under the Fiat, so that didn't look good. I failed to realise at the time that it was oil rather than coolant. It turned out the sump had also been ripped off the bottom of the engine.

We heard a quad bike start up nearby. A local farmer had obviously heard the commotion and decided to come and check out what was going on. We decided to make ourselves scarce, thinking that we had done something dreadfully wrong that would get us into a whole heap of trouble. In fact, we had just had an accident. It was a stupid accident but no one had been hurt. Things could have been far worse.

We manhandled Graham back into the Fiat then all piled in. I started it up and attempted to drive off down the hill. Nothing sounded good and it wasn't steering well. We got about 30 yards and a rattle from the engine turned into a horrendous screeching noise. With the oil gone, the engine was seizing up.

There was clearly no way it was going any further, so we decided to push the car all the way down the hill, across town, and up the other side to get to Graham's house. To help the car roll better, we managed to straighten the drag link a bit and, given that Graham had a broken leg, we let him steer.

I can hardly believe it now but we must have pushed that car for a couple of miles, mainly uphill. I guess four fit young blokes can handle that sort of thing. Thinking back, it was a good thing that there were three guys in the back seat. When the *Dukes of Hazzard* stunt drivers jumped the Dodge Chargers, they carried ballast in the back to stop the weight of the engine tipping the car forward and into the ground nose first. That might well have happened with the Fiat if I had been on my own, without three strapping lads in the back seat.

We left the car in a layby opposite Graham's house. I then had to phone up Mum at stupid o'clock and get her AA membership number so that we could have the thing hauled back to my house. I had a vague tinker with it to try to fix it but it was too far gone.

The Fiat ended its life as another hulk on the driveway outside Mum's house, sitting alongside what was left of my yellow Bug. In the end I weighed it in – took it to the breaker's yard and sold it for scrap. The best thing about that car had really been the engine and with that gone, it made no sense to start on the rest of the repairs that it needed.

* * *

As part of my uni course, we had to do a year out working in industry, and I immediately thought it would be a great idea to go to California. My friend Steve studied electric guitar at 'rock school' (Grove School of Music) and it sounded like the perfect place to 'study': sunshine, surfing, cool cars, and theoretically I would even get paid.

However, one of my lecturers saw through my cunning plan and suggested, quite strongly, that I find something a bit closer to home. California would have to wait. I wasn't too downhearted though, because when I started looking around for something cool and a bit different to do, I realised that they were working on my childhood dream not too far from where we lived.

The lions, tigers, bears and elephants had all been evicted from Windsor Safari Park and were about to be replaced by a miniature Lego world – Legoland. Having been a huge Lego fan as a kid, and a paid-up member of the Lego Club, I was really excited at the prospect of working with Lego's best designers to help create Legoland, and I started gathering the paperwork required for the lengthy application process.

The other thing I thought might be cool was to work in the movies. Special effects seemed like a really good laugh, especially for someone who liked building things. Shepperton Studios wasn't too far from where we lived and I set off one Friday, driving the Fugitive Sand Rail that I had recently finished building, to see if there might be any job vacancies there.

I had a scrapbook of things that I had built and asked at the front desk if I could talk to someone about working in special effects. I was pointed in the direction of a company called Vendetta, who were at that time working on a number of productions. It turned out that they needed someone to help design and build props for a TV series called *Father Ted*.

I showed Jim Francis and Neal Champion from Vendetta FX my scrapbook, pointed to the Fugitive in the car park and explained that I had built that. They mulled it over for all of a few seconds and asked if I could start the next day. I jumped at the chance, and within 24 hours I was working on props for *Father Ted*. Legoland would have been great, but this was even better. And I hadn't even had to fill in any application forms.

Father Ted was set on 'Craggy Island', which was actually several locations in County Clare and elsewhere, with Inisheer in the Aran Islands standing in as Craggy Island in the title sequence. The interior scenes in the Parochial House, where Ted, Dougal and Jack lived, were mainly filmed at the London Studios in central London. However, the special effects team was based at Shepperton, where they figured out how to turn the more outlandish things dreamed up by the script writers into reality.

My first job was working on two cars for an episode called 'Think Fast, Father Ted', in which Father Ted was presented with a new car to raffle in order to raise funds to repair the Parochial House's badly leaking roof. The car was a Rover 200, like the one that had written off my yellow Beetle, and I was about to get my revenge, served nice and cold!

In the storyline, Father Ted and Father Dougal arrived home with the precious car and were admiring it when they spotted a little dink in the wing. Ted couldn't understand how it had got there until Dougal suggested, 'It might have been when you hit that fella on the bike.' Ted decided that a little tap with a hammer would pop the dent out nicely, so he gave it a tap, then another to tidy it up, and ended up pummelling the car to pieces. My job was to turn the Rover from the pristine example that Ted first hits with the hammer into the wreck that he ends up with.

Achieving this was brilliant fun, but no easy task. I tried lots of ways to achieve the desired effect, including using a Kango hammer and blasting the thing with a shotgun, but it turned out that the only way to make it look like someone had gone bananas with a hammer was ... to go bananas with a hammer.

So, each individual dent in the Rover was made by hand with a ball-peen hammer. We started with a car that had already been in a crash and had quite a bit of front-end damage. I had to straighten some of it out and then get to it. I spent a week on it and by the end my wrists were ruined but the effect was great. The car wasn't needed for a while and I suggested that we spray it with a coat of clear lacquer to preserve the areas of bare metal to look 'freshly beaten' but I was told not to bother.

They saved a few quid by not letting me spray it that day but, by the time it came to shipping it out to the set, the car had been parked outdoors for a number of weeks and the exposed metal areas were covered in surface rust. I had to spend a day or two with handfuls of wire wool rubbing down the worst of it before it could go off to filming.

It was all worth it. By the time I had finished with the car, it looked as if it had been parked in an underground car park just before the block of flats above collapsed on top of it. The wheels were fine and the glass was intact but every panel and inch of metal on the car had been beaten, battered and wrenched out of shape. The bonnet and boot lid no longer closed. The bumpers were totally torn out of shape, the wings and doors were a crumpled mess, the pillars and door frames were distorted and the roof looked as if the cast of *Riverdance* had held rehearsals on it.

In the script, when Father Ted managed to get hold of a replacement car to use as a raffle prize, it was wrecked by Father Jack ... which gave me a chance to duff up another Rover. This one had to look OK when Father Ted inspected it from the front, seeing only a damaged number plate. However, when he joined Dougal around the back of the car, the viewers saw that the whole vehicle had been compressed, a concertina of metal all the way from the boot to the rear door. We chopped the car to shorten it and welded it all back together, using a thick aluminium foil (they use it a lot in the movies) to create the concertina bodywork.

Each gag was a new challenge and there were plenty of them on *Father Ted*, including building a dog chucker. This was for an episode where the priests are plagued by rabbits and try to persuade the local dog track owner that it would be a good idea to try greyhound racing with the rabbits chasing the greyhound, for a change. Their efforts were supposed to be so hopeless that the stuffed dog on a trolley fell apart at the end of the dog track.

In the episode 'Hell', drunken old Father Jack's wheelchair had to be seen rolling backwards up 'The Magic Road', supposedly an optical illusion gravity hill where things appear to be able to roll uphill. The scene then required Jack and his wheelchair to be thrown off a cliff at the top.

I thought that the obvious thing to do was to pull the thing up the hill on a cable, however, I was told that it would be too difficult to conceal. So, instead, I used the circular handrail on the wheel as a kind of pulley, wrapped a v-belt around it and then drove that with a car starter motor.

Just under the left hand arm-rest on the inside of the seat, I hid a thin 12v motorbike battery, a solenoid and a belt tensioner with a handle so the actor could slowly let go of the 'clutch' to engage the drive, driving the wheels round at his own pace. I even gave him a dead man's switch on the other arm rest just in case things didn't go to plan. We locked up the front castors so the steering was fixed but adjustable. Once Father Jack was seated in place, his cassock hid the magic. It could do about 20mph backwards on the flat, but going uphill it was suitably slower.

However, when it came to filming, heavy rain had made the 'The Magic Road' rather un-magic and too muddy for the wheels to grip, so they ended up having to use a cable to drag the wheelchair up anyway.

Firing the chair off the cliff was another matter. We used a compressor to build up air pressure in a big receiver tank, and between that tank and a big bore pipe was a kind of pneumatic 'deluge' valve. At the push of a button, this valve released all of the compressed air stored in the tank in one go, out through the pipe. Over that 'exhaust' pipe was slid a second, ever-so-slightly bigger pipe and welded to that was our soon-to-be-flying wheelchair.

When it was time to test it, we set up our contraption outside, giving ourselves as much distance from the workshop as the space in the car park would allow. We aimed our compressed air mortar at the big wooden barn doors and closed them, just in case. The air tank was charged up, a few last checks were made, we aimed, and fired.

With an ear-splitting whoosh, the wheelchair shot backwards through the air and embedded itself deep in the wooden doors. It took us a while to prise it back out, but our machine was definitely ready.

I loved my time working in special effects. To me, TV and movies were magical, and to be allowed to be part of creating even a small part of it was a privilege. I was being paid to play, and being presented with new exciting problems to solve every day. It was a mystery to me how some of the guys who had been in the business for years seemed to have developed a jaded, cynical attitude. I suppose the daily grind eventually gets you down, no matter how fun the work should be.

The Sylvester Stallone movie *Judge Dredd* had recently finished filming when I was at Shepperton and one of the guys I worked with had created special effects props for the movie. He had made the robotic arm for a particularly violent character called Mean Machine. In fact, he had made two arms – one a gleaming, hideous, lethal metal battle prosthetic, and the other an identical soft rubber version used for clobbering Stallone without breaking his bones.

When the filming was finished, both arms were thrown into a skip and set alight. His creation was destroyed just because the filmmakers didn't want anyone getting hold of them and doing something that might drag the film into disrepute. It seemed such a terrible waste.

While I was working at Vendetta, I went to a leaving do. The guy was retiring, having worked his entire career in the industry and I remember toasting him and cheering as he was presented with a watch and some nice words ... and it struck absolute dread into my young heart. I just knew that I never wanted that to be me.

Was that it? You dedicate your entire working life to a job, and at the end of it you just toddle off home to figure out how to fill your days while everything carries on as usual? I knew then that I could never do that. I couldn't commit to this, or to any other 'proper' job, I had to try other things and see what else was out there.

* * *

After my experience in special effects, I thought it would be a good idea to try something completely different. I was a huge fan of the McLaren F1 car and its designer Gordon Murray's work, so I thought I'd give him a call to ask how to get into F1. His son had gone to the same school as me and I figured that was good enough as an introduction, so I just phoned up McLaren and asked to talk to him.

Gordon very kindly took my call and then spent about an hour on the phone with me, at the end of which he told me to go away and design a suspension set-up and bring it in to discuss it with him. What an incredible opportunity!

So, I promptly went off and built a driving sofa instead.

I was taken by the idea of creating a car that was so different from anything else that was around. This train of thought developed into the idea of building a car that looked nothing like a car but was still road legal.

By now, I had quite a few bits of Beetle lying around, and while cutting up a rusty shell to take it to the recyclers, it occurred to me that the back half of a Bug might make a good sofa. I'd had so much fun messing around motorising Father Jack's wheelchair that the notion of a sofa that you could drive really started to grow on me.

How cool would it be to have a car that was a sofa? In fact, wouldn't it be great if you could order a takeaway pizza from your living room, but be outside the restaurant?!

Everyone I mentioned it to smiled, nodded, then told me that it was a *really* stupid idea and that it couldn't possibly be done. Naturally, the more I heard that, the more I was compelled to do it.

By now I had worked at Shepperton and done some work with my uni mate, Dave, at a London design agency called PDD. I had graduated from South Bank Uni but was left with an overwhelming feeling of anti-climax. It seemed like all I had achieved was ownership of a piece of paper. I had a degree. Well, so what? I hadn't done anything that really made a difference to anybody.

That was when I heard about Raleigh International. They are a sustainable development charity whose patrons include Sir Ranulph Fiennes, and they take young people to remote, rural areas in countries such as Costa Rica and Nicaragua to undertake projects like improving access to clean water and helping people to manage their valuable natural resources.

Essentially, it was travelling to another part of the world to help communities and to make a practical difference to real people's lives. It was something that really *meant* something. It was exactly the sort of thing I was looking for and I knew I just had to do it.

I applied to join a diving-based expedition in Belize. I was accepted, but as this was a popular, over-subscribed scheme, there was to be a two-year wait. However, this at least gave me plenty of time to raise the £3,000 I was required to pay into the scheme to take part.

Other Raleigh International volunteers typically organised cake sales, washed cars and did sponsored activities to raise funds. I reckoned that the most obvious thing to do would be to build a driving sofa. I figured that once I had built an awesome piece of roadworthy furniture, I would find a big corporate sponsor grateful to pay loads of cash for all the publicity.

I never did find that sponsor, but the sofa did help me to raise the money for my Belize trip. First of all, though, I had to build the thing ...

I found an old red velour Chesterfield sofa and started to play around with the layout. It quickly became clear to me that using Beetle back ends and existing furniture as the basis for the sofa car design really wasn't going to work. I would end up with an engine hanging out of the back, which would make it just look like a mock sofa sitting on top of a car platform. That simply wasn't good enough.

The idea, the ideal, was that every control and every essential function of the vehicle would be hidden, making the sofa look as un-car-like as possible. It had to look exactly like a sofa, in fact. I wanted to hide all of the obvious car stuff, including the wheels, but was struggling with how to do that using my Beetle bits. Time for another solution.

I was wandering through town one day and spotted a classic Mini parked by the kerb. It struck me that the bonnet seemed about sofa seat height. I took a close look at the car. I realised that it would be a tad rude to start pulling apart a stranger's car on the street to see how it might be reconfigured as a sofa, so started looking for my own in a specialist breakers yard.

I found a rolling front subframe with a low mileage 998cc engine and an automatic gearbox. Good news, one less pedal to hide. Back at the workshop, I immediately realised that having the engine under the seat, as I had first envisaged, would still make the sofa too tall. It would somehow have to fit into the seat back. Drawing it all up on my student copy of 2D Auto CAD, I worked out that I could use the front sub-frame from the Mini at the rear of my sofa and perhaps a front wheel and suspension arm from a Reliant Robin, to hold up the front.

This would give me a wheelbase of two-and-a-half feet and effectively make my sofa a mid-engined sports car. With a really good turning circle. Cool! As it happened, the Mini's rear brake drums were pretty much the same as the front drums, except the rear brake assembly had slightly different back plates to fit the brake shoes and the handbrake mechanism. So, all I had to do to fit a handbrake was take the back plate off a rear brake assembly and modify it to fit my front assembly. The handbrake handle would be hidden under the sofa cushion at the driver's left hand.

Having added the engine 'lump' and the wheel positions to my CAD I then measured a few sofas to get an idea of the range of dimensions I had to play with regarding seat height and depth, back rest height and angle and the most comfortable height and width for the arm rests. A sofa shape was definitely forming but my CAD drawing was looking very boxy and not very comfortable or inviting. It was certainly no Chesterfield, so I rounded off a few corners to make the basic structure look more like it was part of a living room rather than part of an armoured personnel carrier. The result was very Deco 'streamliner' or at least a 1950s interpretation of Art Deco furniture, which was much more like it.

All the way through the design process, I would spend hours discussing the various iterations and bouncing ideas around with Dave. A true kindred spirit, Dave totally got that a sofa you could drive was a hugely sensible idea, and he massively contributed to the design. The outline was done – now, I could start concentrating on positioning the controls.

On encountering the driving sofa, no one should be able to fathom how it was being propelled. It should appear to be pure magic. So, there were to be no pedals in front of the driver – that would be way too obvious – plus, on a sofa you'd surely expect to be able to cruise along with your feet up on the coffee table, like you were still at home? Even better, your mum wouldn't be there to tell you off.

Still, regardless of how silly the sofa was to appear, the most important thing was for it to be structurally and mechanically sound, safe and roadworthy. That's how I always try to do everything; the more frivolous, effortless and ridiculous something seems, the more utterly serious the underlying effort has to be. If you are to put yourself out on a limb and play the fool, it can't fail on a mundane detail.

Having designed the whole thing, and worked out how everything would fit together, I started to build it, welding up the chassis from my two front sub frames and using steel tubing to make a frame onto which the MDF body panels would be attached. There were a lot of angles to cut on the ends of the tubing and they had to be pretty accurate.

Rather than painstakingly mark out each end with a ruler, square and scriber and hack sawing off the waste I developed a quicker solution.

Using my 2D CAD drawing, I printed out each profile at 1:1 scale and then stuck them onto the steel with photo-mount spray glue. Using a 1mm thick cutting disc on an angle grinder saved even more time.

It seemed obvious to Dave and I that we needed to capture and celebrate the kitch look of the British living room from the 1950s to the '70s. The more unfashionable it was, the more it would make people smile. We decided on a leopard-print fur fabric for the sofa covering, a look then typically sported by Bet Lynch in *Coronation Street*.

With the engine hidden inside the now fur-insulated sofa back, and with its original Mini radiator and mechanical fan still attached to the side of the engine block, cooling became a bit of an issue. So, I bought an electric fan and fashioned a metal tube to hold it. I then projected that circle onto the curved surface of the back of the sofa and got busy with the jig saw, inset the tube and welded it in place. In the hot-rod world, setting lights, or in this case a fan and its grill, into the bodywork like this is called Frenching. It was a detail that could not be seen from the front or sides but actually gave the rear purpose and made a world of difference to the coolant temperature.

Originally the rear lights came from a VW Type 25 'Wedge' van – I had a few lying around. Even the rear number plate got a bit of a disguise as a picture hanging at a squiffy angle. There was no wall to put the classic three flying porcelain ducks on to, but I insisted on a standard lamp, technically the rear number plate light. It took a few early morning trips to car boot sales to find the right one and, originally, I attached the whole thing so that it could be removed and repositioned easily when parked up.

I later realised that the brain doesn't need to see all of the lamp to get the full effect so, a few months later, and another Frenched hole, only the top half of the lamp was visible. The coffee table legs went the same way. Originally, I planned to use plumbers pipe bending springs or maybe some rubber tubing, but I decided it looked fine without legs and was far more practical.

With regards to protection during an accident, I had a choice. If the occupants were to be flung from the vehicle as projectiles, they needed helmets. If they were to remain part of the carnage then seatbelts would suffice. I went for retractable belts as they were more easily hidden and the lack of helmets was far more sofa-like. The inertia mechanisms nestled above the TR7 head-light pods on each side of the seat back. In the end, it turned out that when driving in heavy rain, which we have done a lot, a helmet is quite a good idea.

For the fuel tank I used a five-gallon jerry can, hidden under a flap in the nearside/passenger wing. The driver's controls were in the offside wing. I had looked at a number of existing solutions for hand controls but nothing really seemed easily applicable, so I started with the hardest one, the foot brake.

The Mini brake master cylinder sits vertically in the engine bay and its piston is operated from underneath. I mounted that in the front of the wing and then hinged an L-shaped bracket at its corner – the short lever acting on the master cylinder, the long lever to be pulled by the driver

I didn't want an Edwardian style handbrake lever, so I then suspended a round tube under a slot in the top surface of the wing. Over this ran just the right size of square tubing, forming a sliding carriage. To the top of the carriage was attached a tube of metal just smaller than a drinks can and underneath was hinged a long sleeve that slid nicely over the L-shaped lever. Sliding the drinks can back along the top of the wing operated the foot brake. It even had a sprung loaded 'fly-off' latch on top so the brake could be left on for 'parking'.

To be able to operate the throttle with your feet on the table required some kind of knee plunger. This pushed on a paddle inside the wing, which twisted a vertical tube, which in turn swung another lever, which pulled on the throttle cable. It was a little more complicated than necessary to allow the retro-fitting of a subtle foot pedal, just in case the knee operation felt too weird.

Under the leopard print fabric was a line of rocker switches to operate the ignition, starter, lights and standard lamp. The battery lived behind a hinged part of the same wing. Although the Mini engine had an automatic gearbox the gears still needed to be selected from reverse to neutral to one, two, three and drive. Another sliding carriage, this time with a chocolate bar on top, operated a series of ball-jointed levers which eventually pushed and pulled the gear selector cable going into the gearbox. I carved a wooden replacement for the chocolate and glued on a Cadbury's Caramel wrapper, with its appropriate sounding strap-line 'Relax'. Years later I did the same with a Fry's Turkish Delight as a little nod to Cockney rhyming slang – 'Having a Turkish (bath)'/'having a laugh'.

The kidney bean-shaped coffee table out in front sat on a steel frame which was hinged to make it easier to get in and out. It provided physical protection as a front bumper and also psychological protection as a visual break from the tarmac rushing at you when travelling along.

To capture the kitsch, I wanted the surface of the table to be decorated with a mosaic of old coins – very 60s. It hadn't even occurred to me that the various charity shops I was foraging in would expect a small fortune for a jar of old money. So instead, I left bins in local bars and pubs and collected hundreds of brightly coloured bottle tops. Hand washed, and then dried in Mum's oven, I sorted them into colours and arranged them in a flowery rainbow pattern before sealing them in clear resin.

On top of the table lived the original Mini speedometer disguised in an old mantel clock. Initially the speedo cable was routed straight out of the back of the clock and then in through the table, so to cover it up I made a stack of video boxes with themed, punning titles: *Night of the Living Room Dead*, starring Peter Cushion and Sofa Loren, *The Good, the Bad and the Comfy* ...

On either side, the indicators were hidden inside translucent orange flower pots in which grew the finest fake flowers. No self-respecting living room would be complete without a TV, and I found an old 12v camping telly in a second-hand store and wired it to the on-board battery. Getting good reception was easy, just drive around a bit more, although rain was a bit of a problem. I have lost count of how many TVs have been blown up over the years.

Something I love about the movie *Top Secret* is that however many times I watch it, there always seems to be yet another great gag that I haven't noticed before. I wanted the sofa to deliver the same pleasure in all of its detail.

Hiding the steering was perhaps the trickiest challenge. I could have gone with handlebars. In Edwardian times, tillers were common; perhaps even a joystick might have been a scary alternative but in the end, I just considered what you would have on your lap when lounging on a sofa. A newspaper, a book, a bowl of popcorn, a TV dinner on a tray, perhaps a pizza?

That made sense, especially as I figured I could still blag a real pizza pan from my old employers. A medium pizza seemed the perfect size, so I bolted the metal pan on top of a custom-made removeable steering column that was supported through a VW camper van wheel bearing. I would put in a fresh, un-cut, pepperoni pizza every couple of months or so, or every time it got nicked. It was a little terrifying how long the pizzas would last before being consumed by mould, but perhaps even worse was the number of street urchins that must had given themselves food poisoning. (Some years ago, I commissioned a replica which is actually weatherproof and is heavy enough not to take off in a gust when we are overtaking in the outside lane of a motorway.)

Of course, if I wanted my sofa to be road legal, I needed it to pass an MOT test. This was easier said than done. In its original incarnation, it was a three-wheeler. Finding a testing station that would undertake an MOT on a three-wheeler was difficult enough, as they have to have special kit to test the brakes properly. Finding a testing station that would MOT a three-wheeled sofa car was practically impossible.

The testing station where I was a regular customer didn't want to know. I tried other, nearby stations that could handle a three-wheeler and they didn't want to touch me with a barge pole. Clearly, they all feared a subsequent visit from a man from the Ministry of Transport, demanding an explanation for allowing such an abomination onto the road.

Eventually, I took the sofa back to the place that I knew best, and simply pleaded with them to test it. The owner was just about to retire and hand the business over to his son, and his son took the plucky decision to take pity on me. It passed. I got my MOT. The sofa may have looked distinctly odd but it was still roadworthy.

The first time we took the sofa up to London was amazing. It was a night in my life that I will never forget. Driving the sofa is actually quite lovely – the engine burbles and vibrates gently at your back and keeps you warm (it's kind of like sitting on the lap of a purring cat, rather than the other way around). At motorway speeds, the wind noise makes conversations a little challenging (ever tried sticking your head out of the window on the motorway?) but cruising around town that's not an issue and it's a lovely place for a chat.

Dave and I drove up the M3 and started the evening by swinging by a pub near Chelsea to meet a friend. We parked the sofa outside, leaving the standard lamp on, and made our way to the bar to find another friend, Nikki. The pub was heaving, but as we walked in, the sea of people parted, pretty much leaving the bar free for us to be served straightaway. That never happens. It took a moment for us to realise everyone had piled out to look at the bizarre creation parked outside. The peace at the bar didn't last too long though as everyone returned wanting to ask questions. This was no longer a place for a catch-up.

Dave, Nikki and I got back on the sofa and drove into the West End. We rounded Piccadilly Circus and turned towards Leicester Square, passing the Prince of Wales Theatre, where there were hordes of people behind barriers expectantly waiting for the arrival of Prince who was, apparently, about to do an impromptu concert. They were clearly already very excited and, as the first few rows caught a glimpse of us cruising past, they started to clap and cheer and take photos.

Suddenly camera flashes were going off everywhere as the rest of the crowd followed the lead and erupted in exuberant cheering and applause. Mistaken identity or not, it was fun to enjoy the warmth of a crowd and feel a little like rock stars. We continued on our cruise around the busy night-life streets of Chinatown, Covent Garden and, eventually, Soho.

Everywhere we went, people gave us thumbs-ups, smiles and peals of joyous laughter. Tourists were taking photographs, cabbies were smiling and waving, even bus drivers let us pass. We stopped outside Bar Italia, our regular coffee-shop haunt when we came up to London, and the staff came out, grinning, and gave us free coffee. A tramp stood at a distance swearing at us in disbelief for several minutes, then suddenly stopped, came closer, and quietly asked, 'Does it move?'

Somebody else came up, said they were putting on a show and would like us to advertise it, then gave me their details. It was one thing after another. We finished our coffees and crawled up Frith Street through the crowds, to see more sights. A second tramp watched us go past him and then, seemingly very confused, shouted after us, 'Does it move?' Clearly, that was the issue that was concerning London's tramps.

It was an amazing, funny, epic night that will live with me forever. It was wonderful and I felt really emotional and high on the excitement of it all. The whole evening seemed to pass in a lovely warm, fuzzy blur. There was a lot of love for my sofa and I was basking in its reflected glory.

Maybe in a way it felt like vindication – for once, one of my crazy plans had come together and everybody seemed to admire it and think it was brilliant. It made everyone smile, or even laugh out loud. Looking back, I guess it was quite an epiphany in my life. 'I can do this,' I thought. 'I can dream up these extraordinary ideas and make them happen.' I banked that for later too.

The next day, Dave and I decided we had to christen the sofa. We decided that London was definitely its spiritual home, and so we turned to pseudo Cockney rhyming slang to give it its name: the Casual Lofa.

When I drove a living room out of Mum's garage and off down the road, her long-suffering neighbours must have thought that I had finally flipped. However, in actual fact, they should have been grateful for my latest bout of madness, as it became the central focus to my life for a while and resulted in my usual motley collection of half-started and half-finished car projects disappearing from my mum's driveway, improving the look of the neighbourhood no end.

That was only a tiny fraction of the changes my motoring sofa was to usher in. Its effects on my existence were to be way more radical and profound. In fact, the Casual Lofa was to change the course of my life in more ways than I could ever have begun to imagine.

6.

MY EXCELLENT EDDVENTURE

The whole point of the Casual Lofa (and certainly my excuse for creating it) had of course been to raise money for charity so I could take part in my Raleigh International expedition. And, just as I had hoped, as soon as it was finished, my sofa car started getting noticed by the press. Somehow, the *Sun* managed to get the scoop even before it was officially street legal.

For the weeks following, the Casual Lofa was in high demand from newspapers, TV, magazines, all kinds of media. My mission was still to raise money, but without a big sponsor my only option was to charge for each story and appearance.

I didn't quite make it to my £3,000 target before setting off for Belize in the early spring of 1998, but I managed to beg and borrow enough to make up the difference, promising that I would settle any debts once I got back and hit the road on my sofa again. Now, I had to concentrate on preparing for three months in Belize. I had days left.

That preparation had to involve a bit of 'spring cleaning'. Mum needed me to get rid of all the stuff that I had crammed into her garage and crowded onto her driveway. *All of it*. Anything that I couldn't find somewhere to store free of charge simply had to go.

It was fair enough. She hadn't gone all hard-line and zero tolerance on me; she simply needed it all cleared because she had decided to have an extension to the house built while I was gone. The small prefab garage to the side – where I had skinned so many knuckles and spent so much time shuffling under jacked-up jalopies and rummaging around in oily engine bays – was to be demolished. Deep sadness. However, she was replacing it with a bigger garage that would have heating, lights and proper electrics. Deep joy! So, there was a big upside to all of this change, but also a heavy downside.

One of the many project cars that I had to get rid of was my 1968 Plymouth Barracuda fastback. The Barracuda was a gorgeous car that I had bought for all the wrong reasons and kept hold of for all the right ones.

When I was a boy with nothing much to do, I used to cycle all round our local area and had always noticed a big car parked under a tarpaulin in one street not far from our house. The car never seemed to shift or go anywhere and it always intrigued me. I spent ages wondering what it was. I guessed from its size and shape that it was American, probably some kind of muscle car, and I eventually managed to glimpse enough of it peeking out from under the tarpaulin to satisfy myself that it was a dark green Plymouth Barracuda.

It must have been at least ten years after I first spotted the car that I eventually knocked on the house door to ask if it might be for sale. The guy who owned the car said that yes, it might be, and we took a proper look at it together.

It was fantastic. I loved the styling, the long bonnet and the large, sloping rear screen. Like so many muscle cars based on American saloon cars (the Barracuda was based on the more conservative and sedate Valiant), the two doors were long, keeping the lines elegant, with no pillar breaking up the flow between the A pillar at the windscreen and the C pillar at the rear. A long door also meant that it had a wide opening, always a boon for someone of my size and shape.

It had a 318 V8 – that's 318 cubic inches capacity, equivalent to a 5.2 litre engine – and that was the main (wrong) reason that I wanted the car. My plan was to use that engine in a van project. In the end, I couldn't bring myself to do it. The V8 belonged in the Barracuda. Removing it would have been like ripping out the car's soul.

I decided that I should get the car back on the road – the right reason for buying it. It had no MOT but it had been registered in the UK because the owner was able to give me a V5 document. There was a bit of rust and a few problems to be sorted out but it started and the engine ran sweetly.

Naturally, it needed to be de-Americanised for British roads, which meant that I had a few fiddly jobs to do, such as changing the lights. That is often where imported cars, or car projects in general, grind to a halt. Lots of little problems coming one after the other dilute the owner's enthusiasm until the whole car starts to seem like it is more trouble than it is worth. Thankfully, I have never thought like that, it's just that all of the newer projects get in the way!

I drove the Barracuda away and parked it on a friend's driveway, ready to start work on it the very first chance I got. It was a fantastic car but, while my enthusiasm for it was genuine and I was happy to put in the effort, I found that I never quite had the time.

So, sadly, the wonderful Barracuda had to go. I can't remember exactly where it went but I do hope that whoever ended up with it was able to restore it properly. That car really needed to be driven, to be out roaring down the open road, and I shudder at the thought that it might have been weighed in for scrap. That would be a tragedy.

* * *

On childhood holidays, the anticipation of the journey was always a major part of the experience for me. In some ways, I enjoyed the idea of going on a trip more intensely than I actually enjoyed being away. I think I used to get so worked up looking forward to and enjoying the journey that, when we finally got there, it was a little bit of an anti-climax. This wasn't going to be the case this time, though.

The selection process for the trip had involved a bit of an outward-bound hike, some problem-solving activities and a chat to make sure that we weren't going to go bonkers once we were separated from our home environment. I took all of that pretty much in my stride and it whetted my appetite. It also probably helped that, at 26, I was one of the oldest on the trip and as old as some of the expedition leaders, so the people in charge didn't perceive me as a potential problem. I am also a very strong swimmer, which is useful when you are going to be doing a lot of diving.

I have always been confident in the water. On my family holidays in Cornwall, we would be in the sea all day for weeks on end. When I was very small, I remember Dad used to take me to the local swimming pool, usually on a Thursday or a Friday. Afterwards, we would pick up fish and chips for supper on the way home and I would eat them watching *Top of the Pops* on TV.

At my primary school, Rushmoor Independent in Farnborough, I got very good at swimming and became a member of the Rushmoor Royals swimming team. At one point, I was doing nine training sessions every week. Mum would get me up and out early in the morning to swim before school, there would be another session after school and I would also train on weekend mornings. On particularly cold winter mornings, we had to break up the ice that had formed on the swimming pool overnight before jumping in.

I was very serious about it. I swam in galas at weekends and was even tipped as a potential Great Britain Commonwealth Games team member, if I kept it up into my teens. Years later at KES, I set a record time for 50-metre backstroke – a two-length race, which I won by a full length. I know my record stood for many, many years; maybe it's still unbroken?

However, there were downsides to competitive swimming as well. I would get nervous before races, but contrarily, my nervous energy manifested as a sort of lethargy which gave me the overwhelming urge to curl up and go to sleep. I never quite understood that one. Thankfully, any nerves would vanish as soon as I hit the water and it was extremely handy that back-crawl was my fastest stroke, as that meant I actually started most of my races in the water.

Competitive swimming became less of a priority for me once I went to secondary school. It was no longer practical for me to train back home at my old club and my new school couldn't support the same high level of training. In any case, I soon developed other interests. Still, my love of water and swimming in it and under it is what drew me to the expedition in Belize. Learning to scuba dive was something I had wanted to do for many years.

* * *

There were around 115 of us on the trip, with about 40 staff looking after us. We flew to Cancun and were put on coaches into Belize. We stopped for breakfast at Altun Ha, the ruins of an ancient Mayan city. It was magical and awe-inspiring to spend time in a place that had been created some 3,000 years before.

It was incredibly exciting, and right from the start it felt like exactly the kind of adventure I had been hoping to go on.

Camp Oakley was our base camp – basically a handful of shacks centred around a large open cabana and, a suitable distance from everything, a his & hers long-drop toilet. We wrestled up our tents along the edge of a grassy savanna. At night, we had to wear head-torches to find our way to the toilet, and would emerge to find thousands of spiders' eyes glittering back at us, like jewels in the darkness.

The next day we set off deeper into the jungle with a wise old local guide called Winston who trained the SAS in jungle survival. We trekked to a 'camp' under the jungle canopy where the undergrowth had been cleared. We learned how to build a *basha* – a canvas stretcher bed whose parallel poles are supported off the ground by an A-frame at each end and then a tarpaulin and mosquito net suspended above to keep the occupant safe from the rain and 'nature' around them.

Most of the poles were supplied but we were taught how to wield a machete correctly and how to tie a square lash knot, whether using string, rope or vine, to hold everything together. Winston taught us the importance of only chopping down the right trees. The wrong ones had poison sap just under the bark, and if it squirted into your face, it could blind you. He had learned this the hard way and lost sight in one eye.

These new skills were tricky to master in such a short time and, as the evening was drawing in and we were losing the light, to save time I teamed up with some girls in my group and set to work on creating a triple *basha* – I like to think it was in the Mayan temple, pyramid style. I had found a small group of trees that were growing in the perfect formation for my foray into jungle architecture.

We had limited resources, so using the natural A-frames more than once was a boon. Alison, Anna, Bex and Kate's mate, Kate rallied round in a flurry of canvas, poles and square lashings, while I made the most of my height advantage. The inky darkness enveloped us, but not before we had completed our – slightly saggy – des res for the night. It was a triumph.

Not that anybody got much sleep. The jungle creatures came alive and the unfamiliar sounds were so loud. Perhaps the most terrifying were the blood-curdling screeches from the howler monkeys, shaking the trees as they chased each other through the canopy just above us.

The shared experiences made strong and lasting friendships, but just as each group got familiar the organisers would shuffle everyone around. We were always kept just out of our comfort zone, the various roles in each group changing every day so we all got to lead, cook, build and try stuff we never would have dreamed of doing at home. The whole point was for us to grow.

We were all worried about the killer spiders, poisonous 'tommy hammer' snakes, scorpions and vampire bats. So many things were out to get you. Just a week in, I woke up one morning with a big tarantula crawling slowly across my face. I did have a mosquito net but my face was pressed right up against it. There might as well have been nothing between us. I carefully took hold of the netting either side of the spider and catapulted it into the shrubbery.

One morning, I cleverly got dressed inside the protection of my *basha*. I rummaged through my kit bag and unpacked some fresh shorts only to find another deadly bird-eating monster, alive and well, staring back at me with its many eyes. I didn't bother to hang around long enough to find out exactly how angry it was.

Belize used to be a British colony, and there was still a fairly strong British Army presence there to help persuade neighbouring Guatemala that invading wasn't a good idea. The army there went by the name of BATSUB – short for British Army Training Support Unit Belize. They were to be our knights in shining camo; should any of us get into trouble, they would whisk us out of there in one of their helicopters.

The work itself was tough and very interesting. There were a number of different Raleigh projects across the country, including things like archaeological digs and building projects, helping local communities. I was assigned two construction projects, one building a brick corn mill to allow people from in and around the village to make their own flour, and the other constructing a huge wooden school building in the middle of virgin rain forest to be used for research and preservation efforts. Before I did that, though, I was to spend a month taking part in a dive project on a tropical island near the outer barrier reef called Cary Caye.

The Belizean government needed help to make a case to convince farmers that destroying virgin forest to grow cash crops was a really bad idea. Typically, a farmer would cut down an area of trees and jungle and then plant a field of something like pineapples with the hope of making a quick buck. The rains would come and wash both the crop and much of the soil into the river system which, in turn, would drain into the sea, killing off the coral that had taken hundreds of years to grow, ruining the marine ecosystem. They wanted the farmers to focus on preservation and conservation, putting their efforts into eco-tourism, a sustainable and ultimately more profitable business venture.

Our job was to map marked transect strips of the seabed by hand, so someone could then use that information to calibrate a new 3D sonar mapping technology. A transect line was carefully anchored to the seabed between GPS coordinates. We would then dive down with a metre-square frame, place it gently next to the line and map the contents of the frame on our waterproof sketch pads. The idea was to sketch out and note down the different types of coral (alive or dead), turtle grass, sponges, fans, giant clams and the like. Once that square was surveyed, we would reposition the frame to map the next quadrant.

We needed to be quick but accurate, keeping an eye on our air supply as it usually only lasted between 15 and 30 minutes. While we surfaced to refill our tanks and decompress, the next team would descend and carry on the work. We worked like that in shifts, each team normally managing two or three dives per day.

It was vital that we didn't touch or damage the coral and sometimes the swell and currents made it tricky to 'hover' safely over our work. I found that if I added a little extra air to my BCD (buoyancy control device) I could hang upside-down above the quadrant and control my distance from the seabed by swimming with my fins against the positive buoyancy. It might have looked odd but it gave me lots of agility and control.

Once we had finished mapping an entire strip, our marine biologist would collate all of the information, stitching our sketches and notes together. This information was then used to calibrate a 3D sonar scan along the same transect lines. Once the 3D sonar was fine-tuned, regular surveys could be made of the entire bay, building an accurate picture of the changing health of the coral and life on the seabed.

As our time under water and number of dives per day was limited to prevent any cases of 'the bends', there was plenty of time between dives for other jobs, necessary or otherwise.

We were the first inhabitants of the island since it had been blitzed with pesticide some years earlier and so got the job of establishing a camp and getting much of the equipment and stores over to the island from the nearest port. Riding the laden speed boats towards the blue horizon was exhilarating and we felt very intrepid, like modern day Robinson Crusoes, with a little more luggage and a lot less abandonment.

Once we had explored our tropical paradise and established the best place to set up the mess tent, the marine research 'office' and our communal dining area, we formed small groups and cleared areas of undergrowth to form 'dormitories' where we could set up our individual *bashas*. Alison, Gavin, Bex and I built our structures around a well-established tree with amazing bark. For want of the right name we called it a Peely Weely tree, though it turned out the actual name is the gumbo limbo, and the locals called it 'tourist tree' because its bark is red and peeling. We did have to share our dorm with hermit crabs, spiders and other crawly things, but a brush fashioned from palm leaves soon tidied them out of harm's way.

As all of our fresh water had to be imported onto the island, we hung our water bottles from short lengths of string tied to our tarpaulin's guy-ropes at night. The rain or dew would collect on the tarp and run down the rope and be saved in our bottles, fresh from the heavens. Drinking enough water to prevent dehydration has its consequences though and we needed to build an environmentally sound and sanitary toilet.

A constant challenge with the island was the fact that it was essentially a big mound of dead and shattered coral that had collected on the seabed. The coral sand making up the beautiful beachy coves was basically parrot fish poo; the fish peck off grains of coral to get at the fleshy polyps living inside. Most of the soil that had amassed on the island was from decomposed vegetation – palm and coconut leaves from the trees that had washed up as seed pods and taken root many years before.

So the nature of an island like this means that you only have to dig down a foot or so before you hit sea water, which makes burying compostable waste a no-no. We skewered plastic pipes deep into the coral and attached a funnel to the top of each one to make pee pipes that could be used by either sex, but for more substantial business we needed a more substantial solution.

To get us up and running we had to resort to hanging over the 'Captain's Log' – a tree trunk wedged in the sea at the far end of the island. The splashing waves and ocean breeze kept it fresh, but it was a bit too public a convenience.

As we couldn't dig a hole, Nod, our marine biologist procured parts from the mainland so we could put our privy on a pedestal. The plan was to create a long drop by building a big tower for our thunder box to sit on and a big 'punch-bag' shaped tarpaulin sack to dangle from. Nod had ordered a selection of steel tubes from which we would build the scaffold but, alas, he was no engineer.

Unfortunately, no one had considered exactly how the tubes would be bolted together and nothing lined up as it needed to. We only had a few rudimentary hand tools so after a few false starts and bit of struggling we bolted up the most critical joints and literally lashed up the rest with string. Despite this, our tower was surprisingly strong.

We erected our 'towering infernal' on a tiny secluded beach with a wonderful panoramic view of the ocean, framed by palm trees. A perfectly elevated position from which to sit and think. We finally finished our 'restroom with a view' and named it 'Emily', after one of the girls whose eighteenth birthday it was. A great honour indeed.

Our food rations were simple and most of the ingredients were either dried or tinned. All jobs in the camp were rotated to give everyone a chance but that meant the quality and quantity of each meal was a little, well, variable. The first meal we had was organised by the smallest, slightest member of the group, and the bolognaise that was served tasted delicious but the quantity served for 25 was what I would usually consume on my own.

This and the extra exertion meant I lost a lot of weight over that first month, I constantly felt hungry and even on one occasion fainted due to sheer lack of calories ...

I had food on my mind much of the time, so I decided to build a mud and stone oven so we could bake bread. We found an old drain pipe T-piece and used that as a directional chimney to control the internal temperature. It took a while to get going the first time but we did manage to rustle up some excellent fairy cakes for the camp.

As meagre as our rations were, some of the indigenous population felt that they deserved a share of our supplies. It started with a suspicious rustle in the corner of the mess tent and grew to an endless stream of hermit crabs breaking through our various barriers, making off with whatever they could get their grubby little claws on.

Our first solution was to capture them and deposit them on the other side of the island but the next day, we were set upon by a new raiding party – or so it seemed at first. Hermit crabs choose their second-hand homes based on size and comfort, but other than that they don't seem that fussy. We found one or two using plastic film canisters rather than shells. To tell them apart, we started to write names on their homes and found that they would travel considerable distances for the promise of a good meal. Sometimes, they'd return from their banishment overnight and be back on the next morning's raid!

As part of our BSAC training we did a few night dives. This was an amazing experience, particularly when the bigger predators came out to look for a late fish dinner. On the way back to the shore one night, we were swimming along the surface on our backs, looking at the myriad stars, when we counted a spectacular 14 shooting stars. A wonderful way to star gaze.

The magic continued as we walked out of the water and tiny glowing stars appeared in the sand. They turned out to be bioluminescent plankton that had got stranded in the sand and been eaten by shrimp which now glowed as they darted about just under the surface of the water. Further on, in the jungle, we saw the bright green glow from the tails of fireflies zipping back and forth under the canopy.

As part of the island's housekeeping we needed to ensure that a BATSUB helicopter would have a safe place to land should we have an emergency. We found a beach area big enough and got busy clearing away any branches that might get caught in the rotor blades or blown around in their wash. We then did a sweep of the sand, removing all other debris that might get sucked up and cause damage. Once we were ready, the BATSUB flew in to check the landing site and brought with them a few supplies. We even got them to take a great aerial shot of our tropical home.

While doing the clearing, I'd noticed a round depression in the sand at the side of our new helipad. It occurred to me that no tropical island should be without a jacuzzi. So we dug away more of the sand to make a more pronounced tub and then lined the base and sides with soft palm fronds to give it a bit of comfort and protect the tarpaulin liner from any rocks and stones.

We filled the tub with sea water and then buried the excess tarp around the top edge, just as you would with a garden pond. I managed to get someone to buy me a length of garden hose on their trip to the mainland, which I perforated along its length and looped it around the bottom of the jacuzzi, pinning it down with some large smooth rocks. We had a ready supply of compressed air for the dive tanks so I filled up a couple of spare cylinders and vented them through my new hose to create the bubbles. Because the expanding compressed air took heat from the water, I added hot rocks from a fire to keep the temperature comfortable. I'm pretty sure that nowadays, with eco-tourism having taken off in Belize, there are plenty of infinitely better beachfront jacuzzis to be found, but I like to think that mine was the trailblazer!

Early one morning, a few of us went on an errand to the second dive site, based on a desert island a few hours' boat ride north of our own. The weather was gorgeous and the sea was calm. As we weaved our way up through the vast coral reef we stopped at various points to measure water visibility by dropping a weight on a line over the side and counting the markers until we could no longer see the weight.

The island our colleagues called home, Coco Plum, was a huge bank of sand with far less vegetation than our own. It was a different kind of idyllic. It was part of a small cluster of islands and some distance behind it was a huge, almost volcanic-looking outcrop covered in trees. Riding the thermals in big circles high above the vegetation were hundreds of frigate birds, their wings pointed like a tetradactyl; it looked like a scene from *Jurassic Park*.

We traded stories with the other Venturers and learned of their struggles to sleep at night while the neighbouring islands were used as a trading post for 'White Lobster'. Apparently, the locals knew the reef so well they would stealthily race their boats up and down the bay, without navigation lights, delivering the Columbian powdered delicacy to America.

The weather was changing and it was getting late so we hopped into our boat and raced back to our own camp. It was a great feeling, getting the skiff up on the plane and trimming the twin engines until they hummed in unison. Skimming along the surface, gently weaving port and starboard to dodge the coral as we went, we could see flashes of colour beneath the surface as the tropical fish darted away from our shadow.

The sun was getting low in the sky and it was becoming harder to see our route through the coral cities beneath and our chart was not as detailed as it might be. We had to go slower and slower to pick our way through, and then suddenly we had to swerve to avoid a forest of staghorn coral revealed by a dip in the swell. Our evasive action sent us over a huge rock hiding in the dark water, we heard a sickening chink as one of the outboards kicked up out of the water.

We had been very lucky; the propeller was only slightly scuffed – the sacrificial anode had taken the brunt of the impact. We had only lost a small chunk of metal and a bit of time. We adjusted our course and on we went, thankfully Cary Caye was soon visible in the distance.

Weirdly, when we finally made it to our little harbour and unloaded our gear from the boat, I had a strange *Waterworld* moment; even though we had only been on the water for a day I had completely found my sea-legs. So much so that when I stepped onto the beach I lost my balance. I felt nauseous almost immediately – as Kevin Costner's character had said in the film 'the land doesn't move right'. I was that ill that I had to go straight to bed and sleep it off.

At Easter, the whole of Belize goes on holiday, so we did the polite thing and joined the festivities. We packed lightly, piled into our aluminium skiffs and set off for the nearest town on the coast, Placencia. We wondered around for a while and all the activity seemed to be centred around a big palm-thatched bar on the beach. Most of the town was here, even our fellow Venturers from Coco Plum.

The bar was putting on all kinds of competitions to keep the revellers entertained – drinking contests, burger-eating contests, punta dancing competitions – and lots of fun was being had by all. It occurred to me that this was all good fodder for letters back home so when a swimming competition came up I went for it. First swimmer back to the bottle of rum on the beach wins it.

I clambered into a powerful speed boat alongside our marine biologist and two burly Belizeans. The boat took us out away from the beach, then further out, then further out still, until we were struggling to see the beach hut in the distance. The engines were cut and we got ready to race. Before the pilot of the boat had a chance to start the race, one of the Belizeans dived in, soon followed by the second.

'Go!' the driver said, and Nod and I dived in. I was suddenly stricken with the thought that we were already behind some stiff competition and we had to prove our mettle. Was I fit enough to keep up with these guys? We were representing Raleigh. We were representing our country!

I was kicking like a torpedo before I even hit the water. I held my breath and swam as hard as I could. I looked up to see where I was going and to grab some air. No sign of the others. I had to catch them, so I sprinted on.

Years before, I was trained to swim front crawl using bi-lateral breathing during galas so I could see where the competition was during a race, but there was no chance of that among all the waves and sea foam. I got into a rhythm of swimming 25 metres or so and then checking my heading. Every time I looked up, the beach hut was to the left or to the right of where I had been swimming. I was zig-zagging terribly. What a waste of time, I just had to keep pushing.

I zigged and zagged some more and eventually felt the surf change to foam around me as I got near the shore. I touched the bottom with my hand and staggered to my feet to run onto the beach. I could see the bottle of rum only metres away. Somehow, I had made it first. I looked around for the others but they were nowhere to be seen. To my amazement, I had arrived a full ten minutes before anyone else. Mission accomplished, I had some news to write home about. (I was a little peeved that our trip newsletter noted my win as a rather less triumphant ten seconds but, hey, a win is a win and I did get that bottle of rum.)

On our last day on Cary Caye, I decided to take a swim around the island. I had already been snorkelling around the coral at any opportunity, between dives and jobs, but never gone the whole way round. There was just so much to explore. As usual, I put on my three-quarter wet suit, strapped on a small weight belt and a diving knife, and grabbed my fins and mask. Snorkelling is like flying over a magical alien world, there is always something to get distracted by, but this time I had a mission.

I headed towards the strongest currents first, at the windward end of the island that got the brunt of the waves and weather. The swell was strong which made it tricky to navigate the huge forest of fire and staghorn corals. If you took a short cut over a slightly deeper part you risked being dumped onto the poisonous spikes when a wave receded. The corals could easily pierce a wetsuit, and the toxins in the fire coral could leave you with nasty stings.

Having made it through the coral unscathed, I turned a corner to come face to face with a shoal of barracuda. We had been lucky enough to get a few brief glimpses of these legendary fish but this was very special. It was a huge group of these slim, powerful, silver-white torpedoes, many of them around five feet long.

Barracuda are aggressive scavengers and are well equipped with a lot of very sharp teeth. They have a reputation for having a go at divers, but it is a rare occurrence. Thankfully the water was very clear and so I guess they could see that even if I was a 'large predator' I didn't have any prey for them to pick clean. Nor was I likely to have a go at such a big group of them.

We had a moment's stand-off, staring at each other, waiting for the first move, and then we went on our way. I swam through their shoal, they swam past me. It was a great privilege to get so close.

I was now swimming with the waves, so was making good progress, along the least explored length of our island. I was on the look-out for Portuguese man o' war jellyfish; only days before, two of our group had got tangled in some long venomous tentacles which wrapped around them. It was clearly extremely painful and took us a long time to peel them off. The stings left marks, like barbed wire, all over their bodies, though white vinegar from the first aid kit helped de-activate the venom and reduce the pain.

I rounded the leeside and went wide around the shallow coral – and the Captain's Log – and was now on the last leg of the swim. I suddenly felt a deeply disturbing grumbling in my stomach. Something unfortunate was about to happen down below, and it had nothing to do with my barracuda encounter. It had far more to do with the trip to the mainland I had made the day before, when I'd helped re-stock our fresh water and fuel ready for the next group.

Skiff laden with empty containers, we made our way to the small port. Having filled up the fresh water, we had to wait a while for the fuel pump to be free so we went for a mooch about. We had a few items to pick up from the tiny general stores on the dock. It was a dodgy-looking old shack but had everything on our lists. They even sold cheeseburgers. Actual food! Well, probably.

We ordered one each and wolfed them down. It was so good I ordered a second and washed it down with a can of grape flavour soda. It was a delicious idea at the time but now, sealed in my rubber suit, hundreds of metres from the safety of our modern convenience, it wasn't feeling so clever.

I clenched and kicked hard. This was a swimming race I was even more determined to win. Very glad of the extra propulsion from my fins, practically hydroplaning all the way, I finally made it to Emily's beach.

Splashing through the serenity of the gently lapping waves I ran onto the shore, flung off my mask, shed my fins, dropped the weight belt and wrenched off my wetsuit as I sprinted up the beach. I shot up the scaffold ladder and made it, with nano-seconds to spare. The relief was as sublime as the view.

* * *

The last month of our trip was spent building an eco-school-house, deep in the jungle, in the north of Belize near a town called Orange Walk. To get there, we canoed for miles up a river system to a tiny hamlet. The building was big but made entirely out of materials from the surrounding jungle – we used 80,000 palm leaves for the roof alone. It was a lot to build in a month so there were two groups working on it.

As my birthday approached in May I was determined to have a fitting feast and I fixed the rota so that I would be preparing the food that day. We had been living on porridge and rice and beans for nearly three months so I figured we needed to liven things up.

Also, for my birthday breakfast I really wanted some toast and marmalade. We had packets of yeast and plenty of flour, so I made up some dough and placed it in a couple of rectangular, army style canteens to prove. Now, I needed an oven.

I used a pair of our standard issue, catering size, aluminium cooking pots; I put the biggest pot on the camp fire with a couple of inches of sand in the bottom and placed the canteens of proved dough on top of the sand. Over top of those I inverted the slightly smaller pot, its rim nestling nicely in the sand, sealing the oven, keeping most of the smoke away from the bread.

In the middle of our camp was a huge old mango tree that was excellent for climbing – from the top you could see for miles. The mangos wouldn't be ripe for many weeks yet but, although green, they seemed fully formed. I picked an arm-full, chopped them up and added an equal measure of sugar, covered the mixture with water and left them to simmer for a while. It worked! I had created a wonderfully tart 'mangolade' that tasted perfect on our morning toast.

For supper I wanted to create pizzas for everyone, all 45 of us, which meant I needed proper cheese. We couldn't exactly pop out to the local deli so, after a cheeky radio call, the BATSUB came to the rescue and delivered our mozzarella by helicopter! *Grazie mille!*

Of course, no birthday celebration would be complete without a cake. Unfortunately, the lack of fridges in the jungle meant we couldn't store fresh eggs, and we had no chickens. However, I noticed a tub of custard powder, which contained egg, so used that instead. Plenty of cocoa powder made it into a luxurious chocolate cake.

There was so much mix, it took around four hours to bake, and in the end the fire got so hot the bottom of the oven melted away. Thankfully the base of our cake was only slightly charred. My mate Alison started a rumour that the cake had a 'special' ingredient, which of course it didn't, but it added to the evening's merriment as half the camp were convinced they were high.

Most people planned to go travelling through South America after the expedition and my mate Kate and Kate's mate, Kate, asked if I would travel with them. They thought maybe taking a male travelling companion along would be safer than travelling around on their own.

But then, early one morning I had to take a call on a sat phone powered by an old car battery, in one of the villager's huts. Dave had received an enquiry from the organisers of a new live show to be staged at the Silverstone circuit. Did I want to let Jeremy Clarkson and Tiff Needell drive the Casual Lofa in their *Top Gear Live* show? They needed an answer by the end of the day ...

So, should I travel through exotic South America with the two lovely Kates, or head home for a date on my sofa with a middle-aged, denim-clad Luddite? I am afraid to say that it was no contest. Clarkson won hands down.

7.

HIDING BEHIND THE SOFA

Like many kids, I used to hide behind the sofa when *Doctor Who* was on telly. Every boy and girl was scared stupid of at least one *Doctor Who* villain – the Daleks, even though they couldn't climb stairs; the Cybermen, even though they couldn't run; or the weird Sea Devils in their old string vests. They used a hover-craft in the Sea Devils episodes, which was cool, but the scary bits still had me cowering behind the sofa.

Thankfully, most people grow out of that sort of thing eventually. However, as a grown-up, I had a whole new sofa to hide behind.

Actually, I didn't really hide behind it. I was more hiding in plain sight. I had always been a bit gawky and awkward in social situations but, since one fateful evening in my teens, I have frequently been plagued by debilitating anxiety. It began at a performance with the school concert band at a dinner dance in the local village hall.

We had enjoyed a lavish dinner along with the other guests and then, backstage, exuberantly warmed up our instruments before heading up on stage to take our places. It was the same routine we went through every time we gave a concert. However, this time, as I perched on my chair, the expectant hum from the audience made something start building in the pit of my stomach and I knew I had to get out of there – and fast.

As the curtain rose, I sprinted off stage left, saxophone still hanging round my neck, and just made it out of the emergency exit before all the lovely raspberry mousse and fizzy drinks from the dinner made an unwelcome high-velocity reappearance in the car park. I was left completely shell-shocked and terrified of this thing that had taken hold of me and that I couldn't control.

After that night, my anxiety became worse and worse until I was completely paralysed by fear whenever I had to do something where I felt on public display. Eventually, this could even be just going to the shops and bumping into someone I knew. I could eat in the refectory at school but not in a restaurant; I could hold my own and argue any point in class, but I couldn't get up in front of the same class to make a presentation.

The anxiety would build and build and I would feel wretched until – well, I retched. After that, I could carry on. It made no sense, which made me very angry, which of course made it worse. I know now that it was just performance anxiety, my adrenaline building because my subconscious perceived the need to step up to a situation. My body interpreted this as a fight-or-flight situation, making me need to hurl to make myself lighter in readiness for the fight ... or to peg it down the road.

I had always had a twinge of it before my sporting performances such as a swimming gala or a basketball match. But in those cases the adrenaline was put to good use during the physical exertion of the competition, so the feeling passed without any further drama (or puking). But this? This was different.

In the end, my mum introduced me to neuro-linguistic programming (NLP). The course helped me to relax and gave me some tools to deal with the issue. It was a significant help, but to this day I still suffer from anxiety and will not willingly take to any kind of stage.

I have found that what helps before any kind of performance is to a) just accept the inevitable and allow myself to hurl, and b) admit to someone, even the audience, that I am feeling this way. It also helps having someone else on stage with me. This is why I was grateful to so often have Mike by my side on stage, as he revelled in the limelight and was very happy to let me hide in his shadow.

Once I had built the Casual Lofa, I found it sort of became a prop for me in social situations. Naturally, people stared when I was driving around in it, but I persuaded myself that they weren't staring at me, they were gawping at my vehicle. I was just the bloke driving the sofa. The sofa was the centre of attention, not me.

I drove everywhere in the sofa. I took it to parties and gave people rides. The sofa made people smile, it made them laugh and it seemed to make the world a happier place wherever I went. It was a wonderful feeling that something that I had created was beaming a little ray of sunshine into people's lives.

Complete strangers would stop to chat, ask questions and take pictures at traffic lights, or wherever else they could get close. I found I was totally comfortable with this kind of attention. These people didn't know me and they couldn't judge me – well, except for thinking I was nuts for driving this crazy vehicle, and that was a judgement I was completely at ease with.

However, it wasn't just passers-by who wanted to take a closer look. I discovered that other road users could behave a bit erratically when seeing a living room driving towards them, filling their rear-view mirror or overtaking them on the motorway. Everyone does a double take and most burst out laughing.

More worryingly, some people think it's a good idea to fumble for a camera in the glovebox and then come perilously close to grab a photo of the sofa, while others can't tear their eyes away and don't pay proper attention to their driving. Thankfully, the Casual Lofa has never actually caused any accidents, but there must have been a few close shaves.

* * *

The Casual Lofa was soon generating so much attention that I found myself committed to building two more crazy cars.

This all happened because the sofa car caught the eye of a PR company, who asked me if I would drive around London promoting the September 1998 Capital Home Show at Olympia.

All I had to do was attach a few Capital Home Show logos to the sofa, which wasn't a problem. The leopard skin sofa covers could be easily removed. I had to clean them every so often anyway and replacing them with whatever livery an advertiser might want was simple.

The PR company could clearly see the Casual Lofa's potential as a mobile advertising billboard and, pleased with the publicity generated for the Capital Home Show, asked me what kind of vehicle I might like to build next. I hadn't actually thought about that at all, I was still having too much fun enjoying my new adventures with driving art. The first things that popped into my head kept with the current theme – maybe a bed or a bathroom ... ?

The next day, the PR company got in touch again and asked me how much a driving bed or bathroom would cost. I had absolutely no idea, but I told them I would let them know, and sat down to figure it all out.

At this stage, I didn't even know how to go about building a road-legal bed or a bathroom. Yet, the more I thought about the proposition, the more solutions presented themselves to me, and I began to believe that this particular adventure could spin out a lot further. If PR companies were willing to pay for me to build these vehicles, I could carry on making ludicrous cars and, a bit like working on *Father Ted*, get paid for having fun. This could actually be a business ...

The idea for my company, Cummfy Banana, was born. I talked it all through with Dave and he agreed to chuck in his job at the design consultancy, PDD, where he had been working since we left uni, and joined me as a director of Cummfy Banana.

There are a couple of reasons why I called it Cummfy Banana. Back when I was arrested for armed robbery, as if being accused of holding up a petrol station with water pistols instead of real guns wasn't ridiculous enough, a strange rumour started to spread that I had actually used a banana. Also, the yellow beach buggy I built from my first car, apparently at a squint, looked like something cartoon superhero *Bananaman* might drive. The nickname 'Edd the Banana' seemed to stick.

The other reason, of course, is that on a sofa you should be comfortable. Not just comfortable, but *comfy*. In fact, really *cummfy*, which is a cool, squishy and relaxed word and also has six letters, just like banana. I liked the symmetry of that. So, there we were. Cummfy Banana – perfect!

* * *

There are many rules and regulations to which you must adhere if you are building a vehicle that you intend to drive on the public road. Most of them are perfectly sensible, some are a bit questionable, and a few are utterly ridiculous. It's all slightly more complicated nowadays, although even when I built the sofa in the late 1990s there were plenty of hoops to jump through. However, with every vehicle we went on to build, I got better and better at hoop-hurdling.

That, however, did not mean that the car was automatically fit for the road. Passing the MOT was a huge thing, but then

I had to send my pass certificate to the DVLA to apply for a chassis number. Nowadays, we have company status and can assign our own chassis numbers, as we are officially a vehicle manufacturer. This has helped to speed up the process considerably. However, back then I had to apply for a chassis number so that I could then also be assigned a registration number for the number plate. This had to be a 'Q' plate, the kind of plate that identifies a car as having – how can I put this? – questionable origins. Back then, kit cars and other one-offs generally had 'Q' plates, and Q424 ABL nestles beneath the Casual Lofa's coffee table to this day.

Yet even then we were not necessarily road legal. The Casual Lofa still had to be seen to conform to the Construction and Use Regulations. We are very lucky in the UK that we have this set of rules, because written rules are always open to interpretation. In other words, you can wangle your way around them if you know them well enough. You have to always obey the rules, of course, but not necessarily in the conventional manner.

There are sensible rules, like maximum width, height and length, and other things like having sufficient power to take the vehicle up slopes of a certain gradient. From the driver's sitting position, he has to have a certain field of view out of the windscreen. Then there are the rules about headlights. They have to be a certain distance from the centre line of the vehicle and at a certain height. However, the rules didn't say anything about longitudinal location. As far as I was concerned, that meant that they didn't have to go at the front.

When building the Casual Lofa, I decided to put mine halfway along the car, pretty much behind me. On the first iteration, I used the pop-up headlight mechanisms from a TR7, sideways, so they actually popped outwards from the side of the backrest of the sofa. As it turned out, this solution was less than perfect as the lack of weight on the mechanism meant my headlights became very sensitive to bumps in the road, regularly flapping in and out, flashing oncoming traffic, after a speed bump or pothole. The headlights now reside, rather more conventionally, just under the coffee table, at the front of the vehicle.

Another requirement I wanted to get around was an internal rear-view mirror, as there was nowhere practical for it to go. The side mirrors that stuck out from the edges of the coffee table were very effective and reasonably unobtrusive, nestling amongst the plastic leaves of the indicator pot plants. But another, central mirror would be difficult to disguise and I knew I didn't really need it.

Happily, as it turned out, I could get around it. The C&U Regs state that an interior rear-view mirror is not required in a vehicle if it is fitted with a bulkhead that meets with the roof – like in a commercial vehicle or a van. I argued that the seat-back of the sofa was the bulkhead and the top of the sofa body was the roof. Because they touched, I complied with the rule and didn't need the third mirror.

Believe me, I am not given to browsing through the minutiae of government legislation just for fun, but the great thing about rules is that they are instructions on how to get away with things.

As long as you can prove, or successfully argue that you comply, there is nothing to stop you from carrying on with your project. So, I needed to know about the regulations to make sure that what I was building was road legal. In fact, I got to know the rules better than most of those who were tasked with enforcing them – Her Majesty's constabulary.

Most police officers were perfectly pleasant, took a look at the paperwork, wished me luck and sent me on my way. However, others seemed to take a dim view of the Casual Lofa, seeing it as an affront to their ordered reality, and made it their business to remove this blight on normality from the public road. I got stopped on the M3 near Feltham once by a local police officer who clearly hadn't yet gone metric.

'You know that there are rules about what you can drive on the road, do you?' he asked me.

'I do,' I truthfully replied.

'And this contraption conforms to all the rules, does it?'

'It does.'

'These headlights, for example, have to be at least 20 inches high, don't they?'

'They don't.'

'You trying to be clever, son?'

'Not at all. The regulations say that they need to be a minimum of 500 millimetres high, not twenty inches. There is a difference.'

'What are you talking about?'

'About eight millimetres.'

It doesn't pay to get lippy with disgruntled police officers, which I suppose I should really have known. The copper gave me a look that he probably normally used to strike fear into the heart of an East End mobster and started circling the car, clearly thinking, 'Right – I'm going to get him for something.'

However, whatever feature he pointed out, I was able to justify by quoting the relevant regulations. This infuriated the officer even more until, in the end, he dispatched his partner to get a copy of the actual regulations. After a very tense half an hour of this guy looking through sheaves of notes, the partner eventually gave the officer an exasperated shake of the head.

The policeman folded away his notes and said, with as much dignity as he could muster: 'OK, you can go. We have to be somewhere else.' He just couldn't bring himself to admit that he couldn't find anything wrong, and that there was absolutely nothing wrong with me driving around in a mobile sitting-room.

Another officer stopped me on the A316 at Richmond one day. He was so intent on finding something to nick me for that he took the cushions off the sofa and was soon waist-deep in the engine bay, arse in the air, looking for something, anything, to fault. The Casual Lofa was by that time already quite well known as it had been featured all over the press and it just so happened that a journalist from the *Sun*, who had been the first national paper to feature it, came past as the police officer scrutinised my creation. Sensing a good story, she screeched to a halt and took some photos, which ended up becoming an article in next day's paper.

This police officer didn't see the funny side of his backside featuring in the press. Having reluctantly been forced to let me go that time, he clearly made it his mission to get revenge. So, when, some weeks later, his colleagues in the motorway control room radioed him that his nemesis – the leopard-print sofa – was on the back of a trailer travelling northbound on the M25, he flipped.

The guy got in his patrol car and drove out of west London all the way around the M25 to the M1 to pull Dave and me over as we were on our way to Donington racetrack for an event. After giving our rig a fervent once-over, he was virtually frothing at the mouth as he proclaimed that I was in contravention of road safety regulations for having non-matching wheel nuts on my trailer. He gave me a prohibition notice for the trailer, and demanded that I unload my vehicles there and then on the hard shoulder of the motorway.

The policeman let slip that he had 'felt compelled to chase me halfway around the M25', presumably some way out of his jurisdiction. I guess he'd not enjoyed his 15 minutes of infamy and the subsequent ribbing from his colleagues. Clearly, he felt I had made a fool of him, and that needed retribution. The charge was, of course, complete rubbish, but I could tell that this guy could not be provoked any further. I just had to give him a victory.

So, I called Paul Brackley, who saved the day in his van, bringing a spare trailer. Paul returned the 'prohibited' trailer to the workshop; I drove the sofa the rest of the way to Donington.

I had heard the escalating admonishments the police officer was receiving over his radio while he was booking us, so I know he got in trouble for what he had done, but I suppose it must have felt worth it for him to have got his little triumph.

When you are given a prohibition notice, a PG9, you are prohibited from using the vehicle until you rectify any faults that have been found. You may be allowed to drive it away unless the vehicle is deemed to be too dangerous to drive, in which case you have to have it towed from the scene. Once you have dealt with the faults, you have to put the vehicle through an MOT test and take the new certificate to a police station, where you are given another document, a PG10, which overrides the prohibition notice. There were no MOTs on trailers in those days, so I just took the prohibition notice and my trailer to the local police station, to check if I needed to do anything to get the notice lifted. The police just laughed and said it was nonsense.

I suppose I was asking for that sort of thing with the sofa ... and in a strange way, that was the point. I wanted to see how far away I could get from a conventional car and still be within the regulations. It seemed the answer was a considerable distance.

The police attention could get a little trying, though. My record for the number of times our Cummfy Banana vehicles were stopped was 27 times in a week and once, in London, 19 times in one day. Nineteen times! Even if each stop lasted only about 10 minutes, that was still more than three hours wasted standing around aimlessly at the side of the road.

Nowadays, we have the IVA (Individual Vehicle Approval) test that has to be carried out at an approved IVA test centre. It's a bit like an MOT, except it's far more involved and is designed to ensure that your car complies with Construction & Use Regulations. Once you have an IVA certificate, your car is basically considered to be brand new and 'Q' plates are not required. Neither are today's police officers itching to pull me over when I'm driving the sofa. They tend to recognise the car, and me, and are more likely to smile and wave, just like everyone else.

* * *

Dave and I worked out what it might cost to build the bed and the bathroom and phoned up the PR company to let them know. I can't remember how much it was now, it certainly wasn't an extortionate amount, but when I told them the price there was silence on the other end of the phone. This was followed by: 'Um, we don't have that kind of budget.' This was to become a very familiar phrase for me over the ensuing years.

Dave and I really wanted to get these projects off the ground, though. Eventually, we did a deal with the PR company whereby we would build and keep the vehicles and they would then hire all three of them from us for the duration of the event they were to be promoting. That arrangement seemed to work for them, and it certainly did for us.

The bed and bathroom were 'commissioned' to promote the 1999 Ideal Home Show at Earls Court, from 18 March to 11 April. The cars would have to be ready way before the start of the show in order to generate interest prior to the opening. With Christmas fast approaching, that only gave us a few weeks. Luckily, I have always loved a challenge.

Having agreed that they wanted a driving bed and bathroom, the PR company then had a problem with visualising the concept. They asked me: what were these things actually going to look like? The obvious answer was that they would look like a bed and a bathroom, but I created a rough 2D CAD drawing to show them the bed.

Unfortunately, that only seemed to confuse matters. They reported back that it just looked like an engineering drawing of ... a bed. 'What were they expecting?' I wondered. 'The Eiffel Tower?' Dave and I did a bit of head scratching and then managed to find a way for them to (in PR agency jargon) 'visualise the concept more effectively'.

The staff of a local bed shop were a bit surprised one Saturday morning when a very tall bloke wearing a flying helmet and goggles came in with his friend for an impromptu photo-shoot. Having found a suitable bed, I sat on it with my arms outstretched, holding a steering wheel I had brought with me, while Dave took lots of photographs. We made sure to bid the staff a cheery farewell on our way out. It seemed only polite.

Dave found a suitable street scene in a magazine and borrowed an early copy of Photoshop to graft the two images together, creating a visual of me in a 'bed car' driving in traffic around Trafalgar Square in London. That did the trick. *Now* the PR people could see what we were about, and we rolled up our sleeves to get on with it.

The first thing we needed, of course, was an actual bed on which we could base the car. It had to look pretty spectacular. A standard single bed with a velour headboard really wasn't what we were after. So, I contacted a local company called Ducal, who were totally up for the idea of a driving bed. They manufactured an ideal wooden four-poster bed and they agreed to supply us with one free of charge.

Ducal didn't just do that, they custom-built a Cummfy King Size version with modified bed posts to our specific requirements. They split the wood for the posts along their length and then routed out a channel in each half. Once the posts had been glued back together they used a wood lathe to give the posts the shapely curves that appeared on the standard piece of furniture. The only difference was that these posts now had a hollow heart.

We were thus able to run steel tubes inside the posts to give the bed structure the extra strength that it would need. Being able to hide this part of the chassis inside the woodwork meant that we didn't spoil the look of the bed with unsightly steel supports. Ducal did us proud and their logo was deservedly slapped on the foot of the bed.

By now, I was very familiar with the shortening of Beetle chassis to make beach buggies but for the bed I opted for a bit of an 'upgrade'. The VW Type 3 followed the same basic design of a Type 1 with a body shell bolted to a floorpan, but was longer and wider and came with better brakes. I found an old VW Squareback as the donor for our new bed. As with many classics now, particularly because of its rarity, it would be considered sacrilege to pillage such a vehicle but back then they were less desirable than the Bug, so much cheaper.

The Squareback also had a rear-mounted, air-cooled, 1600cc 'Pancake' or 'Suitcase' engine which was much flatter than those found in a Beetle. This was ideal. Having less engine protruding into the mattress area was going to make our conversion quicker, easier and hopefully more bed-like.

As with the T1 Beetle and T2 Camper before it, the Type 3 is a rear wheel drive. The transaxle (a gearbox and rear differential combined) sits just in front of the engine and the hand brake and gear lever nestle in the tunnel in front of that. Initially, the gear selector required quite a wide stir to find a gear but the addition of two 'quickshifter' kits reduced the exaggerated fumbling under the covers.

The removable front and rear sub-frames of the Type 3 and the fact that it came with front disc brakes as standard made it a perfect choice. I removed the body from the chassis and then cut away a metre-long section from the middle of the floorpan. The Type 3 chassis has parallel sides and a big radius at each end, as opposed to the Beetle's tapered floorpan, making it much easier to shorten. I then made up a structural steel frame that tied the sub-frames together and defined the outer dimensions of the bed. The wooden (and now steel-reinforced) four-poster bed was then attached to the steel chassis frame.

I know that all sounds quick and easy but, while the actual engineering is not complicated, it all took time. Naturally, we also had to make sure that the engine was in good nick, sort out the ancillaries, brakes and electrics, and ensure that everything conformed to the Construction & Use regulations. It all seemed to take longer than expected, and as they weren't about to delay the opening of the Ideal Home Show on our behalf, we had to burn a lot of midnight oil.

The bed had to have a mattress, of course, but we couldn't have the driver and passengers simply sitting on top. They had to have proper seats, secured to the chassis, and seatbelts if we were to be road-legal. And to make the bed look authentic, the occupants had to look like they were *in* bed, not attached to one.

So, the seats had to nestle inside the mattress. We had neither the time nor the resources to design and commission a custom-built mattress, so we got hold of a standard sprung mattress and attacked it with a 9" angle grinder.

The angle grinder was ideal for slicing through the steel springs and frame but less than ideal for the fabric and wadding, which smoked out the workshop and kept catching fire. We ended up with a mattress with a slot missing that could accommodate bucket seats for the driver and two passengers, who would all sit comfortably reclined under the duvet.

To keep the occupants of the bed in the correct position in relation to the duvet and pillows, which was vital if we were to make the bed look good on the road, I opted to keep the seats in a fixed position. There was no need, after all, for the passengers to be able to adjust their seats forwards or backwards, but we did have to be able to accommodate drivers of varying heights and leg lengths.

If the seat wasn't going to move, then the pedals had to do so instead. So, I used an 'upcycled' adjustable seat base to create a pedal cluster that could be shifted forwards or backwards in a second slot cut out of the mattress. Moveable-pedal set-ups are common in sporty kit cars and even supercars like the Ford GT, so it seemed like a good option for our, ahem, sports bed.

We needed to ensure that we had the right sort of range of adjustment. Luckily, we had the ideal way to ensure the car was comfortable for drivers of all heights. We had two living ergonomes, one from either end of the anthropometric scale – myself for the farthest setting, and Paul for the nearest. Paul had kindly been helping me with the build on weekends and the odd evening. Next to me on TV he might just look far away from the camera, but actually he has a fairly diminutive stature, ideal as a racing driver and for giving me a hand.

We now had the passengers safely, and comfortably, installed, the mechanicals bolted on and the engine plumbed in. It was time to try it out to see how it handled.

Because the VW engine hangs off the rear of the transaxle, and we wanted to keep the motor out of sight beneath the bed, the rear wheels of our bed were a bit nearer the centre of the car than I would have liked, resulting in a very short wheelbase.

This feature gave the bed, like the sofa, a brilliantly tight turning circle but, with all that weight hanging out the back, it also had a dreadful tendency to pop a wheelie. That was fantastic fun at first, but it was clear that something had to be done to keep the front end on the ground. Apart from anything else, I couldn't steer if the front wheels were up in the air!

The obvious solution was to add a weight or counter-balance at the front of the vehicle. I suppose I could have fabricated some kind of frame to support a selection of concrete slabs, dumbbell weights, or even a trough arrangement into which we could have poured concrete, but there wasn't much space between the front beam axle suspension and the front of the bed.

I reckoned that a big chunk of iron or steel – denser, and so more compact for the same weight – would work better, and so I trotted off to the scrapyard in Blackbushe (who I think had forgiven me for the fire by now). As a regular and valued customer – I weighed in an old rusty camper or two pretty much every week – they were happy to let me rummage around for something suitable.

It didn't take too long before I spotted the perfect thing. The scrapyard has a conveyor belt that carries bits of old cars, knackered washing machines and all manner of white goods and other steel products through a series of rotating metal jaws. These chew up the steel and flatten it out. Each tooth or hammer of the jaws runs on a huge rotating crankshaft and is lifted up and down, squishing the steel into smaller and smaller pieces.

That sort of work takes its toll on the hammers eventually and, sitting neglected by a pile of other scrap I found an old, worn-out, metal-crushing scrap hammer. It was a single lump of iron weighing around 200 kilos, roughly the weight of three people ... well, providing that they haven't grown to anything like my size.

It wasn't the easiest of things to move, but we got it into the van and back to the workshop, where I was able to secure it to the chassis, out of sight behind the baseboard of the bed. The bed wasn't quite as quick with the hammer in place, but the handling was vastly improved, the front wheels stayed firmly on the ground, and the extra weight didn't seem to affect the drivability too much.

Now, in addition to getting its paperwork in order, all that we had to do was to dress the car to make it look as if it had come straight from your bedroom. We added a fabric valance below the wooden rail of the bed frame, which not only looked the part but also helped to hide the running gear. There were matching curtains at the head of the bed, and another fabric trim hanging from the top rails that ran between the bedposts.

The only things that didn't look as if they belonged on a bed were the headlights, side lights and indicators at the front, and the rear lights on the back of the headboard. There was no doing without them, of course, but the curtains and fabric skirts softened the look of the car to such an extent that the lights were barely noticeable. The fabric looked fantastic flowing and flapping in the wind when the bed sailed down a road.

Keeping the curtains and skirt attached proved a little trickier than we had anticipated because they were apt to rip themselves free, but keeping the pillows attached to the bed was harder still. There are probably still a few stray pillows lying in roadside ditches all over the place. The duvet also had to be fastened in place, and it helped to keep the bed's occupants warm as they hurtled along in the fresh air.

The fuel tank lived at the front of the bed somewhere under the feet of the passengers. We managed to hide the fuel filler neck inside a repurposed rubber hot water bottle on the foot of the duvet. Details like that don't take long and I believe they are worth the effort.

We called the bed 'Street Sleeper' – a term that is normally used for a stealthily souped-up car that looks completely factory stock but has something a bit special under the bonnet to make it go faster. I like a good pun, and the name stuck to the car far better than its pillows did. Driving along the road, the Street Sleeper gave passers-by as big a surprise as the Casual Lofa, and Dave and I were mightily pleased with our latest creation.

This was no time to rest smugly on our laurels, though. Now we had to crack on with the mobile bathroom. Because we couldn't ignore the fact that the date of the Ideal Home Show was drawing alarmingly close ...

8.

PUTTING ON A SHOW

When I was at boarding school, the first- and second-year boys were housed in Queen Mary House, before moving up to the senior houses in our third year. QMH was originally built in 1867 as a convalescent home and later used as a home for the mentally ill before being bought by the school. It is an impressive and ornate three-storey building and has been used as a location for a number of TV and film dramas, including Agatha Christie's *Poirot*.

The upper two floors housed dormitories and bathrooms but there were only a couple of baths in the bathroom area, which was mainly devoted to showers. After outdoor activities like football and hockey matches during the cold winter months – and in particular, following a gruelling cross country run – we'd fight over the bathtubs and the chance to warm up and relax our tired muscles.

I have never seen the appeal of the painful, sweaty pursuit of running but, being so much longer in the leg than most of my peers, I could lope along and do pretty well in most races. Cross-country, though, was a different matter entirely.

Spending time in the countryside, mud-plugging through puddles and across fields is great fun when you're bouncing around in a decent 4x4 off-roader, but not so much when it's just you, out in the cold and rain, in your PE kit and a pair of trainers and up to your knees in sludge. It was just something we had to do. Despite my considerable misgivings, I found that if I put the effort in and pushed through the pain, I could usually finish somewhere in the top three and therefore guarantee myself a prized place in a relaxing hot bath.

Annoyingly, though, my consistently good performance won me a place in the school's cross-country team! Competing in external races, however, did come with the benefit of being allowed to stay out afterwards with Mum for an afternoon tea in a local pub.

Food also became my motivation to excel in Mrs Lucas' maths lessons. For a year, her maths class was the last lesson before lunch. She would finish the class by asking us some very convoluted mental arithmetic problems; whoever finished each problem first was let out of class early and got to be first in the queue for lunch, if they ran fast enough. Getting first choice of the best food drove me to become very good at mental arithmetic. It's amazing what you can do when you have the right incentive.

Racing for a post-cross-country bath was just one of the many bathroom battles that went on at our school. We also turned the bathrooms into a competitive slide area by blocking the drain holes in the floor with loo roll and turning on all the showers. We could have up to an inch or so of water covering the whole vinyl floor, making it a superb area for sliding barefoot, or bare-arsed.

While we were doing our waterpark thing all that water had to go somewhere. As it wasn't down the drain, it ended up going through the floor into the ceiling of the room below. Leaks from our water rink turned the plaster damp and filled the light fittings. Ironically, some years later, I got a summer job at the school tearing down that very ceiling so that it could be replaced. Oops.

For the third year I moved to Edward House – named for the school's founder Edward VI rather than in anticipation of my tenure. Here, it certainly paid to keep your wits about you if you wanted to avoid being punished by a senior for a misdemeanour around lights out. Usually, that meant having toothpaste squirted up your nose. It may not sound too bad but, believe me, depending on the brand, the toothpaste sticks to the inside of your nose and the mintiness burns like crazy.

Aquafresh, the stripy, red, white and blue stuff, was the worst: the white part was particularly sticky and the blue and red part contained mucho-minty sting. Colgate Blue minty gel was the least punishing as the blue gel was neither too caustic nor very adherent. This meant that as soon as the punisher had got back to their own business, you could easily jettison the offending paste into the sink with a hefty nasal blast. On the bright side, the world smelled incredibly fresh and clean for hours afterwards.

Talking of bathrooms, well baths anyway; we had spent so long constructing our road-going bed that when it was time to create our driving bathroom, we had days rather than weeks left. I say 'we' because I was not working on my own.

Dave and I had a brilliant bunch of friends. Back then, before proper jobs, mortgages and families came along, they had plenty of time on their hands, and were happy to bail us out when we were really up against it. My friend Alison once said that she relied on us for her yearly adrenaline rush. Whether it was a big push towards a build deadline or racing crazy cars at the Nürburgring during a truck grand prix, it was always frantic and always a lot of fun. In fact, another of our mates, Jim, met his wife Beth while driving the bathroom at a wacky race event at Donington Park, where her uncle was driving a motorised shed.

I had a pretty strong idea of how the bathroom would work – in my head. Now I just needed to translate that into a driving reality. It seemed obvious that the driver should sit on the toilet – the place in a bathroom where you would expect to see someone sitting. The bath could go alongside the toilet and would be the ideal place for a brave passenger, or two. Which suggested to me that some kind of motorbike-and-sidecar arrangement would make the most appropriate base, so I started thinking about a suitable bike.

I quickly came to the conclusion that we should be looking for an automatic scooter because I wanted as few controls as possible. It would mean not so many components to hide and, as the rider would have fewer driving actions to worry about, would make it look like they weren't driving at all.

I went to a local bike shop to ask their advice on suitable donor bikes or scooters. The bloke there suggested a much-unloved 1980s commuter scooter called the Honda Spacey, but I was concerned its 125cc engine wouldn't be up to hauling the extra weight of the bathroom and would be hopelessly sluggish. The bike expert had a think and then remembered there was a 250cc version. This was the one for me.

I told him that I wanted a 250cc, but the guy lugubriously assured me that he absolutely never got them in, and they were rarer than hen's teeth. I gave him my phone number, just in case, and went home a little disheartened.

The very next morning, he rang me. 'I can't believe it, but someone came in this morning wanting to sell his gran's old scooter – a Honda Spacey 250!' I was back at his shop with a wedge of cash within the hour.

The 250cc scooter could do 75mph, with very quick acceleration, so that granny must have liked a brisk ride. Whoever she was, Biker Gran had just saved my bacon.

Taking a proper look at the Honda Spacey, I was pleasantly surprised at the overall size of the thing. It looked big enough for our purposes, being reasonably long with its feet-forward riding position. It was a luxury cruiser, *Easy Rider* kind of scooter, with a heavily padded saddle, a high windshield and everything boxed in by moulded plastic panels. It looked a bit 'Gerry Anderson', like some kind of hoverbike out of *Thunderbirds*.

I set about stripping it all down. There was plastic cowling everywhere, all designed to make it look comfortable and civilized for its target market. Thankfully, it all came apart quite easily. Once I had the bike down to its bare bones, it was time to ponder how to actually turn it into a bathroom.

Well, we certainly needed a bath to create the sidecar. I wanted a Victorian roll-top bath, but an iron-and-enamel bath would have been far too heavy. So instead, I needed to find a more lightweight plastic or fibreglass one.

A good bath of this kind would have two fibreglass skins bonded together with a kind of plaster of Paris to give it weight and make it feel comforting and substantial. It also helps to retain the heat, allowing you to wallow in the bath a lot longer before the water goes cold.

Well, that was what you would want in a *proper* bath, but we managed to find a really cheap, pretty crappy quality, GRP one, which was superlight – perfect for us. I wanted the bath to hover just above the ground at 'bath height' but without any sign of the third wheel needed for a side car. My solution was to support the side car with a trailer suspension unit, positioned in the centre of the bath.

To accommodate the diameter of the wheel while keeping the bath at the right height I had to cut a hole in the bottom of the bath so the wheel could poke halfway through. This then meant that the passenger's genitalia would be perilously close to the spinning wheel. To protect against unwanted abrasion, we placed a steel hoop over the wheel, welded it to the sidecar's chassis and then clad it with white plastic. (In a later iteration, we would opt for an unadulterated bath with the wheel on the outside, hidden by white panelling and strategically draped towels that looked as if they had been casually slung there but were actually glued in place.)

The driver had to be sitting on a toilet, which had to go where the Spacey's padded seat was, or he wouldn't be aligned with the handlebars. We couldn't use a porcelain toilet because it would be too heavy and, frankly, too breakable for a roadgoing vehicle. So, in keeping with the Victorian styling, we went for a wooden 'throne' style toilet. Which had the added benefit of having more space for the engine than your average U-bend.

I got a toilet seat, took some measurements and made up a steel frame, which we stylishly concealed with MDF decorated to look like ornate panelling. Your great-granddad would have been proud to go about his business on a throne like this.

I also wanted to make sure that everything we did to the bike was reversible. I used existing brackets so that everything could easily be taken apart and the original panels put back in place. It meant we could turn the powerplant in our toilet back into a Honda Spacey, should we need to. And if something went wrong with the bike, we could bolt in another Spacey 250cc (well, if we could find one).

While I was welding up the chassis and framework for the loo, others were working on painting the MDF. They painted a dark colour over a lighter colour so that the surface could be combed to reveal the paint underneath and give us that luxurious hardwood grain effect. With a couple of coats of varnish, our throne toilet soon looked, well, fit for a king.

Steering was via the hand basin. Obviously. We managed to find a cheap, vacuum-formed vanity sink that was perfect. It was light and simple to attach to the bike handlebars, concealing their basic structure and the fact that they were still being used to control the vehicle.

I put a sheet of blue Perspex over the top of the sink to make it look like it was full of water. You could see the speedometer hidden inside. We installed a pair of chrome taps and used shaving mirrors for wing mirrors. A rubber duck on the sink and a soap-on-a-rope dangled off a tap and made it look lived in.

Immediately behind the driver was a Thomas Crapper-style toilet cistern, the kind that you would normally see on the wall above the toilet to provide the reliable flushing power of gravity. The cistern's pull chain was for the horn and inside the cistern itself nestled the Honda's original fuel tank.

To hide the front wheel and suspension forks, I needed something else commonly found in a bathroom. I went for a good-quality wicker laundry basket as it needed to be robust and structurally stiff enough to take the abuse it was likely to get on the road. I turned it upside down and cut a big slot in the back, running up from the open end, so that it would fit over the wheel and fit either side of the bike frame.

We cut one side off a couple of lengths of box section steel, to make a U, and then clamped the loose ends of the wicker together. The headlamp lit up the road ahead from under the wicker lid. This was attached to the back of the inverted basket, so it could flap in the wind as we went along.

To make everything look more homely and less roadworthy we attached towels to the wicker and added socks, dodgy lingerie and big pants so they were spilling out of an overflowing laundry basket. As usual, everything was about trying to cover up the bike so that you couldn't see how it all worked. The more things flapping in the wind, the better.

The bath had proper Victorian-style mixer taps with a hand shower, and then sprouting out of the top of those stretched some copper piping which ran up to a large overhead shower rose. A loofah and shower cap completed the picture. We named it 'Bog Standard', simply because it wasn't.

Getting our attention-grabbing cars ready in time for the launch was hard enough, but we'd taken on a second task that took an even bigger effort and more late nights to complete.

As part-payment for the hire of the promotional vehicles, we had accepted a 'valuable' advertisement in the show brochure and a large stand space in one of the Earl's Court exhibition halls, as Dave and I felt this was a great chance to drum up more business. All we needed to do was create a stand to display Cummfy Banana's offering. So, our army of friends got busy again.

While I was away in Belize, the majority of the work on Mum's new extension had been finished. That summer I turned into a builder's labourer to help finish it off and, by the time we were working on preparing for Earls Court, Dave had actually moved into our new spare room. Having given up his day job, and being unlikely to see any salary from Cummfy Banana for a while, he saved on rent by living with us.

Dave and I were thus using the spare room and the garage as Cummfy Banana's temporary head office. The neighbours, who had been relieved to see the decomposing metallic clutter of my car detritus disappear, before having to endure builders' vans and cement mixers, must now have thought that some kind of hippy art commune had moved in.

We needed to create a shell-scheme for our exhibition stand that would entice the droves of potential clients who would surely flock onto our stand to engage with the world of Cummfy Banana. So we set up a kind of production line on the driveway to build and decorate panels and backdrops and cut carpet and fake grass for a street scene with the Casual Lofa at the centre.

Just as things were coming together, we were told that the exhibitor behind us at the show had pulled out and that we could use their space, too. There was clearly only one thing for it. We made more panelled walls, painted them with blackboard paint and rushed out to buy a cheap pub pool table and a job lot of chalk. *Voilà* – one instant games room to relax in!

Dave's dad ran a sign-making business so our company logo looked very professional on the panels, especially as Dave had designed a bespoke company font for us. It was one of the first things that he did when he came on board, and he made a brilliant job of it. It all looked fantastic.

Those few months were frantic but exciting times. We had no idea where it would all lead to but we were certainly putting in the hours. If hard work and effort were the only key to success, we were nailing it. Of course, in actual fact, our business proposition at that time was pretty hopeless. We weren't charging nearly enough to build the vehicles and we were utterly reliant on the huge goodwill of our friends to get things done. Still, we were all having a laugh. And for me, this was to be the beginning of a working life in which sleep deprivation played a major part.

Early on the morning that we had to set up at the show, having worked all night yet again, we loaded up Paul's Sprinter and my trailer with our freshly made wall panels, the pool table and rolls of carpet. Those turned out to be a fantastically comfortable place to sleep and I snored gently in the back of the van while Paul battled the commuter traffic into London and Earl's Court.

There were a few celebrities promoting the show, as is usually the way with these things. *Countdown*'s Carol Vorderman, Laurence Llewelyn-Bowen from *Changing Rooms* and Charlie Dimmock from *Ground Force* posed in the vehicles before we took to the busy streets of London.

With three mad vehicles on the road, we needed three drivers, so Dave, Paul and I cruised around town in the Casual Lofa, Street Sleeper and Bog Standard. On these excursions, we learned an awful lot about London and its layout. We weren't given any specific instructions as to where we should drive, it was more a case of: 'Make sure you are seen by as many people as possible.' So, we followed the crowds and turned into streets we had never seen before just to find out where they went, pottering along wherever there were people to take notice.

Everywhere we went, there was a whole load of smiling and waving and countless tourists taking photographs. We were definitely getting the Ideal Home Show noticed. Even better, we knew that we were getting ourselves noticed, too.

However, the fact that the three of us were out driving meant that there was no one left to man the stand. Once again, our loyal friends and family stepped forward to pitch in. They worked on a rota basis, taking turns to stand around, hand out leaflets and field enquiries. When it was quiet, they just grabbed a game of pool.

Thinking about it, our neighbours at the exhibition must have been just as incredulous as Mum's neighbours at home. As our vehicles were out on the road for the entire duration of the show, we had nothing on our stand. The whole point of paying a fortune for a stand at a trade show like that is to show off your wares, and we were exhibiting absolutely diddly-squat.

To increase our fellow exhibitors' bemusement, we not only had an empty stand, but also had another one that we appeared to be using purely to loaf around playing pool. It must have looked like an arrogant waste of space and resources. Yet thankfully, our stand was fairly busy and attracted plenty of visitors curious to find out who this strangely-spelt Cummfy Banana were, and what they actually did.

So, we spent the start of the exhibition week driving our crazy vehicles all over London and chatting to whoever we could about Cummfy Banana. In the hall, we had a few genuine enquiries – and one particular visitor who recognised straight away exactly what we were all about.

Tim Mitchell saw the potential in our young, naïve antics. Tim had recently taken over as MD of the European distributor of an Israeli company called Friendly Robotics. They had made one of the world's first robotic lawn mowers, and they needed to let green-fingered suburbanites who wanted a cutting-edge garden know all about it.

Tim looked at our sofa, bed and bathroom cars and instantly figured that a giant, street-legal robotic lawn mower would be his perfect marketing tool. On an empty stand, in the depths of Earls Court, Cummfy Banana had secured its next crazy commission ...

9.

CUMMFY LIFE

As Cummfy Banana became a reality and we'd started work on our first commission for the bed and bathroom vehicles, Dave and I had started looking around for a workshop to call our own.

The solution came from my old mate Mark at Just Kampers. He had recently relocated to a farm in Odiham and had spied some available space for us in the old grain store. It wasn't exactly a state-of-the-art workshop with workbenches, heating, lighting and all mod cons (it didn't even have a toilet). But we were off Mum's drive, much to the neighbours' delight, and we only had to share it with an owl, and the mice and rats that normally made up its lunch.

The premises were rough and ready; we were in one end of the barn next to a huge open 'corridor' than ran straight through the building from one side to the other. An icy wind whipped through almost constantly. In the floor of this open space, as wide as the doors, was a massive grille, used to dump huge loads of grain into a big hopper below. It was like a magnet for our tools. I spent many hours climbing down into this void and fishing around in the piles of old grain to recover sockets and spanners. Still, at least there were no neighbours that we could upset by hammering, banging and welding till late into the night, and it was in a lovely countryside location.

We could see RAF Odiham from just outside the barn. When Chinook helicopters flew overhead, we would hear the distinctive double-whump noise of their twin rotors – and we could tell which way they were going by which part of the barn was vibrating. Otherwise it was a quiet area and eventually we'd film the first series of *Wheeler Dealers* there.

However, by the time we got started on our next commission – 'Robomow', for Tim Mitchell from Friendly Robotics – the harvest was approaching, so we had to vacate the grain store. And, anyway, surely this new project deserved a workshop with a less tool-hungry floor. So we moved further up the farm into a unit next to Just Kampers, inside a huge old Second World War hangar.

We had given Tim the most hopeless of quotes for the job – way below what we should have charged, inadvertently leaving off so many items. We ended up losing money on the project, but that was down to our total lack of experience in costing things properly rather than anything to do with Tim.

What Tim was expecting from us was something like an old A-Class Merc tarted up with a vinyl wrap and a few bits bolted on. We were aiming rather higher and created a bespoke vehicle, an exact, road legal replica of his robotic lawnmower. It definitely exceeded his expectations.

To make the body, Dave digitised the lawnmower product and then scaled up those measurements to the size of a car. We then made patterns for the fibreglass moulds and had a huge fibreglass bodyshell made. I built a chassis to connect the original suspension sub-frames and running gear from an old Rover and I incorporated the braking mechanism and the accelerator into the steering wheel mounted on a telescopic column – pull back to brake, push forward to accelerate. It made driving it quite surreal, especially in reverse.

At the back were huge replica fibreglass wheels that would roll along on grass, and hidden inboard of the body were the car's actual wheels. A hydraulic system allowed the body to be lifted up to provide plenty of ground clearance on the road, and dropped down so the big wheels could make contact with the grass when on display at an event.

We wanted our creation to look just like the real thing, so our bodyshell had no windows, which obviously made visibility rather difficult. For use at shows and events we fitted a video camera to the front and rear of the body, connected to a video monitor. They flicked from forward to reverse with the gears but didn't give a fantastic view of the outside world. It was a little like playing a computer game with only one life! When parked up, the monitor could be raised up to display promotional films.

Driving on the road needed a more conventional solution. The Robomow mower had a removeable battery box, so we followed that design cue and turned that lid into a roof for the driver and passenger. They sat on a pair of car seats that were mounted to a frame on a custom-made scissor-lift, also powered by the same hydraulic system. At the push of a button, the two occupants with their seats and telescopic steering column would be lifted up out of the body giving them full view of the road. In hindsight, we probably should not have made it possible to do this while driving along but Tim loved showing off the feature and it certainly got people's attention.

Tim was so impressed with our work that he commissioned a second promotional vehicle for the launch of the new model. Robomow 2 was a much cooler shape than the first and its styling really lent itself to becoming a car. We upped the budget and designed a much more capable vehicle than the first, applying everything we had learned from the first machine.

This time, the big wheels at the back really *did* work and they were turned by a mighty Rover V8 engine, which Tim used to great effect on a display lap at Le Mans, stealing all the thunder from a new clutch of 'Lexi', and enthusiastically spinning off into the grass around the Nürburgring F1 Circuit at a truck grand prix.

We used much the same process to create our scale replica fibreglass bodyshell, but this time we added windows. Once the 2D sketch was in CAD, we realised we needed to stretch the body if we were going to fit the engine and driver under the same shell. I stretched the profile on my CAD software to fit and then added an extra side window for visibility.

Even when next to the original product, the extra length didn't make it any less recognisable, but it did give the car more purpose and balance visually. The windscreen was made of three identical pieces that were sections of a cone. We had them specially made by Pilkington AGR and to save time we had the body laid up in fibreglass simultaneously. This meant we had to create two sets of identical wooden patterns for the screens, one set for the inside curve and one set for the outside. Thankfully the laminated screens fitted nicely onto the body.

To hide the screen, and to make the car's profile more like the product, we had some special perforated vinyl printed to match the yellow gel coat. On the inside it was black so you could see through it as if the glass had a slight tint, but someone on the outside would only see the colour and not notice the occupants inside the car.

The new Robomow model had a sculpted yellow body sat on top of a black rubber bumper so we used that split line in our version. We welded up a tubular frame and bonded it to the inside of the light-weight body, hinging it at the front, below the screens, creating a big 'clamshell' opening. We used a pair of big gas struts to open the lid and a pair of bigger pneumatic rams to close it. The compressed air also operated the latches to open the clamshell via a radio key-fob.

It looked super cool and was great if you lost the car in a big car park, but was annoying in the rain and could be awkward in covered car parks. I once got stuck in an underground hotel car park in Sweden – the roof opened and immediately hit a concrete beam. It was the only parking space available, so I had to shuffle back and forth until I found the gap between two of the roof beams and then had the ignominy of slithering out of a tiny gap onto the floor. The car did seem a little less cool at that moment.

The projects kept coming. For the launch of the 2002 *Scooby-Doo* movie, we were asked to convert three LDV vans to look like the Mystery Machine and create a 'beach shack' pop-up stage for a promo tour. LDV had somehow managed to persuade Warner Bros. that their latest Sherpa van looked just like the Mystery Machine. It really didn't, but, by the time we had finished with the three that we were asked to convert, they came pretty close.

British Leyland (LDV Group used to be called Leyland DAF Vans) had, some years before, finally upgraded from round headlights to the more modern square ones, so we had to spend ages customising the front of the van to revert back to round headlights to match the Mystery Machine. The vans also needed to have American Racing spoked alloy wheels, which meant fabricating hub-plate adaptors to take a different wheel stud arrangement and flaring the wheel arches to accommodate the wider wheels and new offset. We added neon lights underneath the van to give it a, well, *mysterious glow*, and then delivered them painted in a basic blue colour ready for a vinyl company to add the decals that looked like the Mystery Machine's custom-paint job.

For the mobile beach hut scene, we were given a low-loader transit van and a look at the film's style bible – the brief was to create a kind of mobile beach shack stage for 'Scooby Karaoke'. We painted some six-inch diameter aluminium tubing to look like driftwood logs and hinged the longest one at the base of the cab. The onboard winch was then used to lift it up at an angle and two shorter supporting lengths were fixed to this spine as an A-frame. A branded canvas was then draped over the structure and we built a rustic wooden staircase for people to get onto the back of the truck. Using the winch meant the whole structure could be erected and packed up in minutes.

For Haagen-Dazs we created the 'Pamper Wagon'; a large six wheeled creation with cylindrical pods for ice cream sampling at the rear and a luxuriant red seat at the front covered with a pillowed surrey top, from which hung a 'pleasure sensing' device (a 1950s-style beauty salon hairdryer covered in flashing lights). It didn't need to be street legal as the ice cream samples were to be given out in places like train stations and shopping centres, so we based it on an electric golf buggy.

It looked like a cross between a lunar buggy and a hareem from *One Thousand and One Nights*. A customer would sit in the sumptuous chair and be given ice cream samples. A 'scientist' would 'measure' their increasing pleasure with each new sample until an 'orgasmic' siren would be set off by their peak of ice cream bliss.

One of our great enduring clients was Innocent Drinks. Richard, one of the founders, got in touch when they were just starting out and looking for a bit of direction on how to create a visual presence for their brand. They wanted a look for a new sampling van. We worked with them for weeks until we finally hit upon the idea of grass and daisies. It seemed to encapsulate perfectly the brand's ethos and the origins and honesty of their product.

We created the 'Innocent DGV', or 'Dancing Grass Van'. It was an ice cream van covered in fake grass and daisies, which 'danced' on hydraulic rams to music (funk seemed to work best) blasting out from giant aluminium megaphones on the roof. Whitby Morrison, the ice-cream-van people, built a body onto our customised VW van chassis and, in the meantime, we commissioned a special drinks fridge to go in the back so it could be taken to events for sampling and promotions.

Innocent loved the DGV so much that they commissioned several matching delivery vans. We did various 'Grass & Daisy' projects with Innocent over the years. However, as they grew, their ever-expanding delivery fleet became beyond our level of supply. We were a creative agency, not a fleet management agency, but it was lovely to see that the concept lived on even though we no longer made the delivery vans.

As Cummfy Banana progressed in a flurry of new creations and promotional campaigns, life moved on. Dave married Mandy and moved to Australia, where he got a job as a product designer at Breville. It was the end of an era and I was sad to see my best friend move so far away but I was happy for him. He's got a lovely family and a lovely life and he did make it back to be best man at my wedding to Imogen. I carried on building weird vehicles and soon, my creations increasingly got me embroiled in other projects, like event appearances, TV shows and breaking records ...

* * *

Dave was still here, though, when I first got clocked doing 87mph in the sofa at a wacky races event at Donington racetrack. Neither Dave nor I had ever been to the circuit before that and had no idea what the course looked like for this one lap race.

Thankfully, on the back of the entry ticket was an image of the track, so Dave 'navigated' while I chucked the sofa into each corner in hot pursuit of a win. We were up against stiff competition: an orange, a skip, a banana-covered 2CV, even an armchair. We got off to a great start and were in the lead at the first corner. Dave's navigation helped us stay on course, and after a few more bends we thought we had it in the bag.

As we climbed up the incline of Starkey's Bridge we were being caught up by the skip. By the time we were going round Coppice, the skip's turbo was on song and it was steaming past us. But then puffs of tyre smoke came off the tyres as the skip slewed right through the Esses – clearly he had overcooked it.

We seized the opportunity and nipped up the inside, but there wasn't enough room. Fighting for a piece of track, our weight was thrown over to the right as we shot left into Goddards. The sofa was up on two wheels – the front wheel in the centre perilously close to the rumble strip. The skip went wide as we thundered onto the finish straight. We were neck and neck but we had kept our momentum. Down came the chequered flag and we were just in front, a classic win!

As the official timing results from the race were published, Craig Glenday, the editor of *Guinness World Records*, said that the Casual Lofa's 87mph made it the world's fastest furniture and asked my permission to print it in their next book. I was delighted, I had been a big fan of the brothers McWhirter and their work since I was a kid. It was also the start of a long-standing relationship with *GWR* and provided me with yet another focus for my creations – they now also needed to be record breakers.

Following on from the success of the sofa, came the fastest bathroom, fastest office, fastest shed, fastest bed and largest motorised shopping trolley ...

The world's fastest office, 'the Hot Desk', was a commission from the Business Environment Group. Actually, they just wanted a driving office, it was me who wanted to set a record in it, but they were perfectly happy for me to do so. Like our previous creations I wanted the office to be as un-car-like as possible.

I had an idea that an L-shaped office car might be just the job, so we agreed on the style of furniture and I got busy on CAD. After accommodating the big 'lumps' such as the engine and rear suspension, the design ended up looking something between a '?' and a '7'.

As with many of our vehicles, it was based on the much unloved but super-reliable, automatic Rover 100 (CVT) with the front subframe and engine at the front of the vehicle in front of the driver. This was housed in a steel box frame which extended asymmetrically down the right-hand side (off-side) of the vehicle to the back wheels. The rear subframe was narrowed so the back wheels sat either side of the centre line of the car but would fit inside a square 'cupboard'.

On top of the structure sat a number of curvaceous, specially made, laminated table tops to give the vehicle a modern 'ergonomic' feel. The shapes of the furniture units and cupboards beneath echoed the same curves. The tubular steel structure was hidden with wooden 'doors' and, on their rear, or outside of the vehicle, we used brushed aluminium modesty panels.

At the back of the car, on top of the rear wheels, was a round meeting desk with three reinforced office chairs (with seat-belts) attached to the steel structure. We held many meetings there over the years, but it was tricky to take notes with the wind tugging at your papers at 60mph. Just forward of the meeting table was a pair of cupboards for storing promotional material and next to that we incorporated a classic office water cooler which dispensed hot and cold water.

I was adamant that there should be no floor inside the '?' shape as I didn't want the Hot Desk to be like a carnival float. It worked a treat because it really messed with your head when it was parked up in the street.

The driver sat at the desk at the front, in a reinforced office chair, with the brake and throttle pedal by his feet and the gear selector just under the desk. To the driver's right we fixed an open briefcase which housed the dashboard, ignition switch and one of the rear-view mirrors. Next to that was a desk tidy-type pen holder, in which was hidden the headlight and indicator stalk, and some pens, obviously.

On the desk in front of the driver was a working wireless keyboard that was also a racing style quick-release steering wheel. To steer, you rotated the whole keyboard (rather than pressing 'Z' & 'X'). Just left of the driver we bolted down a working computer monitor, connected to a laptop hidden under the desk. For a really authentic office experience we installed a Premicell, which was a device often used on oil rigs and the like, and used a SIM card to give landline-style phones connectivity, so we could make calls and use the internet while driving along. Well, actually we couldn't – although back then it was still legal to use a phone, you could only use a monitor for 'information that pertained to the journey' or something similar.

We finished the Hot Desk the day before the big press launch in Hanover Square in London. I drove the office up from the workshop in Odiham, about 50 miles away, just to shake out any bugs. Paul followed in a van and trailer, just in case.

While the press assembled, we waited tentatively in a nearby side street. Any minute now we would be given the signal to start our engine, and make our triumphant entrance around the square for our big reveal in front of the press. As we waited nervously, a cycle courier on a racing bike sped up to us, a little too close, but grinning broadly. As he passed, he stuck his arse in the air and cracked off a rectal bark as loud as a claxon.

Being the well brought up, mature gentlemen we are, Paul and I looked at each other indignantly and then started to snigger. We appreciated the tension relief, it was quite funny, but then, when he looked back to see if we'd heard him, the courier cycled straight into a parked car and ended up sprawled on the bonnet. We fell about laughing ... and, to give him his due, so did he.

Press shoot complete, I then started on the next leg of our tour, down to BE's head office in Basingstoke. After more photos and a cup of tea, we sped on down to the coast and then on to Cannes in the South of France. My friend from school, Sue, is fluent in French and kindly wrote me a note for the *gendarmes* explaining that our jalopy was fully street legal and insured.

At virtually every motorway toll check point, we were pulled over by the police. Some wanted to look at the paperwork, but nearly all of them just wanted to take a few photos. We got as far as Lyon, where one official didn't want to let us go any further. He thought we would hold up traffic by going too slowly. We cured him of that delusion by sitting him down, strapping him in and zooming around at up to 80mph before whisking him back to his own office. He waved us through and we made it to Cannes just in time.

* * *

The Hot Desk build was another fraught, sleep-deprived project. Largely because I had agreed to build a giant shopping trolley for the London premiere of the movie *Jackass* at the same time. I roped in the help of as many friends as I could muster, and even made a new one.

An article about Cummfy's creations had been published in *Top Gear* magazine, and I had said that I was always looking for staff, asking people to get in touch. My now great friend, Neil, came down for an 'interview' in the middle of the mayhem of my double project and promptly left, three days later. He just walked in, got stuck in and within minutes it was like he'd been part of the crew forever.

Years later we rebranded the Hot Desk for a job with Vodafone. We were to be the lead car on a set of parade laps up and down Regent's Street for an F1 event in London. It was an amazing privilege to lead the pack. There were thousands of spectators lining the streets who had been waiting for a glimpse for hours. Once everything was ready, the course was opened and off we went.

The streets were so packed that after our run, there was no way into the pits at Waterloo Place. We were parked in a service lane just up from the action giving us a prime viewing position inside the plastic barriers.

The F1 cars screamed up and down the street and a few did donuts in front of the pits. The excited crowds, now at fever pitch, cheered, waved and whistled, but things were beginning to get out of hand. The hordes of spectators were now heaving, cordons bursting at the seams.

The last car ran down the hill and the organisers immediately closed the barriers behind them. There was now no way back into safety for us. A shiver went down my spine as I noticed an unnerving noise behind us. I turned round as the noise became a din; like some twisted zombie movie, hundreds of manic fans were sprinting down the street towards the pits, and we were in their path.

There was nowhere for us to go, and nothing for us to do but get into the crash position and wait for the mob to clamber through us. It was insane. As the rush went past, people were jumping onto the barriers and onto our table top to just to get through.

After a few minutes, the rush turned back into a fervent stream and then someone seemingly recognised me. I was wearing red racing overalls, just like an oversize racing driver. Somebody thought I *was* a racing driver and asked for my autograph, and then another, then another. Some four hours later we were finally able to pick our way through the mountains of empty drinks bottles and rubbish and make our way home.

* * *

The shopping trolley was a beast. It was over 11 feet tall and nearly 10 feet long. I'd digitised an actual shopping trolley that I borrowed from a Sainsbury's car park late one evening and scaled it up in my CAD.

It was going to take a lot of stainless steel and a lot of welding to get this beast together. I ordered about £60,000-worth of polished stainless-steel tubing and had it cut and bent to shape. Over the course of one stupidly sleep-deprived 24 hours, Paul, Tall Paul, and I welded up the basket. It remained suspended from the forks of my forklift until we could weld it onto the chassis we had yet to assemble.

The original trolley chassis used rounded rectangular section tubing but I couldn't find anything like it in the size we needed. I found a brilliant company in Woking who were able to bend two lots of 100mm diameter stainless steel tubing into the inverted 'tick' shape we needed. They did an amazing job of getting one frame to fit perfectly on top of the other. Tall Paul, Paul's friend from the drag-racing scene who often worked with us, then TIG-welded a joining strip to cover the valley between the tubes, all the way along their length, on both sides. The chassis looked just like the real thing.

The trolley needed to run on castors, but without the random wobble. I went for slightly bigger wheels than were strictly to scale as I needed to fit in brake discs and callipers. A company in Birmingham built me a set of custom three-piece alloy wheels that looked just right and I then shod them with the lowest profile tyres that would fit. My plan was to drive the rear wheels with electric motors and pivot the front casters for the steering. The hard bit was actually making the castor frames themselves. For strength and scale they were made from 16mm stainless steel plate that was first guillotined into shape and then each half folded 90 degrees.

Half a mile away from the industrial estate where the steel was being cut and bent we could hear, and feel the ground shaking. We could see rings forming on the surface of our cups of tea as the giant press pounded them into shape. T. Rex is coming. Trolley-shoppus rex, that is; the king of the shopping trolleys.

The batteries and controllers were hidden in the basket as boxes of 'Horn Flakes' and cans of 'EddBull', and we made giant branded bags for promotional jobs. The driver steered the trolley from the toddler seat. As I am about three times the height of a toddler, the trolley's 3:1 scale seemed ideal.

Driving it was a bit hairy, especially round corners, as you sat so high up in a spindly structure which bucked and swayed as you went along. *Guinness World Records* bestowed our creation with the honour of being 'the world's largest motorised shopping trolley'.

* * *

That was not a record I thought to undertake again, but I did break several of my own records over the years – well, someone had to. I'm not particularly driven to collect accolades, but it is a lot of fun. I built a fire-breathing V8 version of the Lofa, the 'Chaise Lunge' with the engine hanging out the back. We broke my 'fastest furniture' record at Bruntingthorpe, in the pouring rain, at 92mph. I say we, as, though I was in the car as an instructor, the honour of driving the car and getting the record had been auctioned off for charity.

In Bog Standard, I broke the record for the fastest bathroom in 2011, at just over 42 mph and then broke that record at Car Fest some years later, in 2018, at a hair under 50mph.

This was, in a way, thanks to Freddie Flintoff. A couple of years earlier, Freddie attempted to break 24 records in 24 hours for charity and *GWR* asked me if I would help him set a new mobile bathroom record on Bog Standard. So, I fitted nitrous oxide to the toilet to give it that extra bit of boost.

Unfortunately, when it came to setting the record, it was the crack of dawn on an icy, freezing morning and the nitrous froze, depriving us of our boost. Again, I say 'we' as, strangely, Freddie did not want to drive a souped-up toilet, and reclined in the bath as a passenger. The weight of the two of us strapping, magnificent athletes was just a little too much for the Bog Standard to accelerate past its previous peak speed.

Still, I left the nitrous set-up on the toilet and we got another chance at the record at Car Fest North at Bolesworth. This time on a gorgeous hot day, with the gas expanded to full pressure, I was 'flushed' to finally get the result.

* * *

In 2003, the Cummfy Banana HQ was invaded by a bunch of dubious media types who set up camp and stayed for about three months. Having said yes to appearing in *Wheeler Dealers*, the production company convinced me it would be a great idea to use my workshop for the filming. It certainly was. They got a rent-free studio space and full access to all our tools and equipment.

Still, I thought it was a good idea too. If I was on site, I would be able to continue to run Cummfy Banana alongside filming the show. Well, that was a naïve and ridiculous notion but I'm nothing if not stubborn, so the way I made it work was by simply not sleeping.

When the film crew packed up and went home for the evening, I'd have a quick bite to eat and start work on Cummfy business, or any prep work that needed to be done on the *WD* cars, sometimes carrying on until the film crew returned in the morning to start a new day's filming. I probably managed to get through that first series by sleeping about two to three nights per week – not to be recommended.

During Series 1, we refurbished a 1985 Saab 900 Turbo and one of the jobs I had to do was swap the nasty beige cloth interior for a much more stylish black/grey leather interior. It was a big job and we finished filming the removal on a Friday afternoon. The crew left at about 5pm to beat the rush hour traffic and I got straight on with an urgent Cummfy project. I had a lot to get done before the end of the weekend and I finished up working all the way through to Monday morning without any sleep. I was just finishing things off when the crew turned up to continue filming the Saab job.

We got cracking with replacing the interior and managed to finish filming the job by the end of the day. I was extremely tired when we started, but absolutely shattered by the last piece to camera. No wonder, when the show came out, the Saab interior change seemed such an exhausting job!

I was able to make that kind of ridiculousness work for the first couple of series as the filming of a series only took about three months, leaving me with the rest of the year to recover – both myself and my business. However, as the show grew, both in popularity and with the number of episodes per series, I could no longer keep it up.

I had to start turning business away; I no longer had the time to design and build new vehicles and started to wind down that part of the business. We kept up with the vehicle hire for a couple of years but, after a while, even that became too much of a distraction and Cummfy Banana had to be mothballed.

The Casual Lofa on the M3 motorway hard shoulder, being scruti-nised by the boys in blue. Again.

Building Bog Standard in a grain store. By this time tomorrow, this will be a driving bathroom – honest. All we need is a bit of smoke and (steamed up) mirrors.

e. Paul and
ll Paul.
let you
rk out
o's who.

ust finishing
the wiring,
en it's time
o make the
d... and the
bed frame.

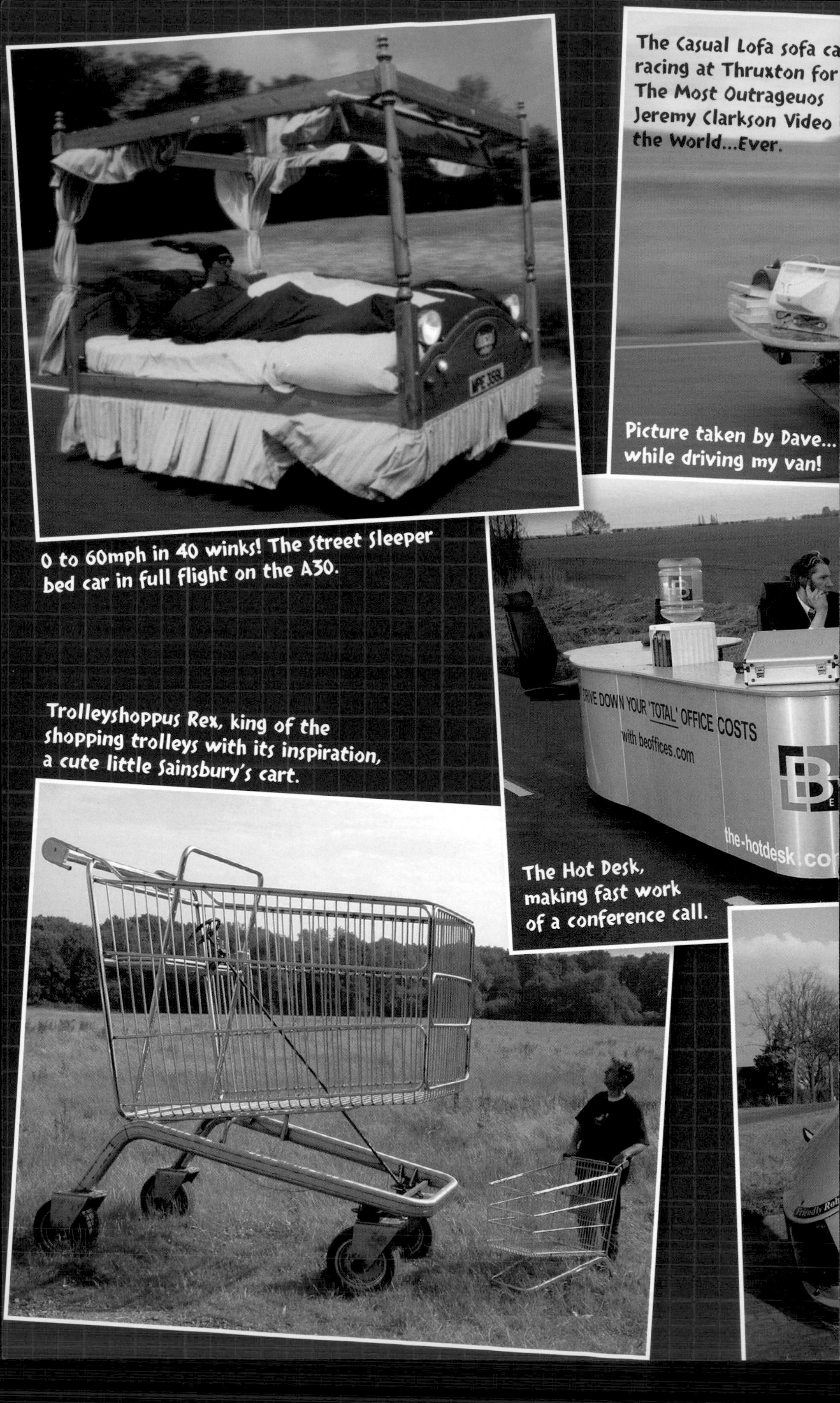
DRIVE DOWN YOUR 'TOTAL' OFFICE COSTS
with beoffices.com

The Casual Lofa sofa ca
racing at Thruxton for
The Most Outrageuos
Jeremy Clarkson Video
the World...Ever.

Picture taken by Dave...
while driving my van!

0 to 60mph in 40 winks! The Street Sleeper bed car in full flight on the A30.

Trolleyshoppus Rex, king of the shopping trolleys with its inspiration, a cute little Sainsbury's cart.

The Hot Desk, making fast work of a conference call.

Dave and I taking the Bog Standard bathroom bike out for a brisk morning constitutional.

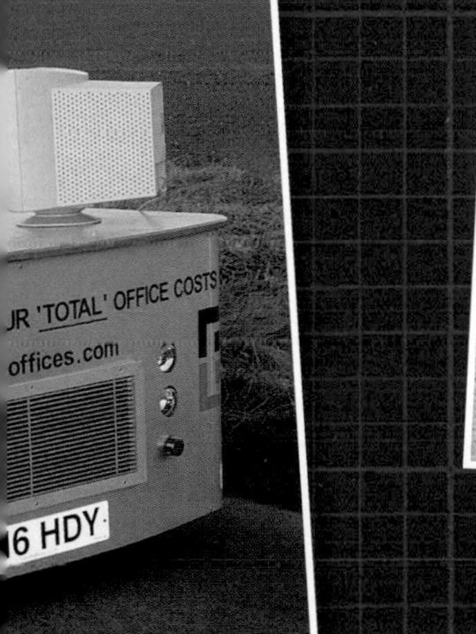

Gone To Speed, shed car, built for 'Shed of the Year Awards' seen here resting at Calais before filming.

The V8 Robomow (2), ...cess, the whole body hinges open.

My lovely Reliant Robi
double-engined dra
racer being crushed fo
losing the race on BBC2'
Panic Mechanics

This all-metal dune buggy started out as a hearse.

The Slow Boat From China gravity racer built from a cut-down diesel Escort van and sporting a handy extra one tonne of flushable potential energy.

Filling up from a fire engine took hours!

ing in the Humdinga after
ting my first speed record
Coniston Water.

My K7 silver star for establishing
two new records.

On the Thames, in the
WD Amphicar with Mike,
before the gearbox blew.

Joining the swim-ins and
arsing about in the Aquada
at Amphib 17 on the River
Thames.

Fun stuff to do in the workshop; welding, grinding, twisted fire-starting!

Making a sunny day into a pea-souper, soda-blasting the filler and rust off the Amphicar.

On the start-line of the London to Brighton Veteran Run in our 1903 Darr
just look at those bright LED headlam

vith Paul on Wheeler Dealers rescuing yet
nother automotive treasure, the 1954
hevy Stepside pick-up truck from Series 8.

Original oil lamp insert for the Lucas 'King of the Road' coach lamps.

3D SolidWorks model of the X12 chainsaw engine that powered...

...my ride-on hand-tool-powered buggy for Silverline tools.

My SolidWorks model of the new LED insert, ready to 3D-print.

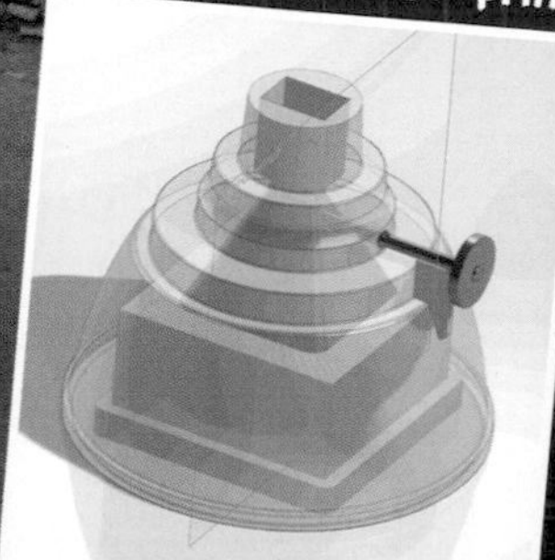

Imogen and I love casually loafing about town in the Casual Lofa.

A box of brand new water-jet profiled adaptor plates for the ice cream van's electric conversion.

Diesel free ice cream anyone?

But, at the time of *Wheeler Dealers* Series 1, Cummfy had three guys working full-time, as well as a couple of admin people, and more on a temporary basis when necessary. So, knowing now how much space the filming took up, we decided to bid our hangar HQ goodbye and find bigger premises. *Wheeler Dealers* Series 2 was filmed at another barn workshop, in a place called Cockpole Green, on Crazies Hill, near Henley-on-Thames.

When the series was finished, Attaboy asked me to be part of another of their productions: the Shed of the Year Awards. The show was celebrating the unique fondness we British have for our garden sheds and exploring some of the more extraordinary examples lurking in gardens around the nation.

So, it seemed obvious that I should build a street-legal shed and drive it to France to see what the French made of it all. I suppose it's fair to say I got a mixed response. Most people seemed to think I had escaped from secure containment but at least one lady really took a shine to me, I think I might have pulled. Later, I set another Guinness World Record for the 'world's fastest shed', so all in all a successful project, I think.

We made three *Wheeler Dealers* series at Cockpole Green but it was an awkward arrangement. The Cummfy Banana builds and the TV cars often tended to be happening at the same time, and there are only so many hours in the day (and night). We simply couldn't have our guys hammering and banging away in the background while I was trying to explain the intricacies of refurbishing a carburettor to camera.

Thankfully, not all of the filming was done at the barn. Mike's pieces – where he was filmed sourcing the cars, his test drives before buying, the test drives once I had fixed them and his sales deals – were all filmed elsewhere. In fact, Mike wasn't really seen around the workshop very often during the early years. For the first two series, the part where he was handing over the cars to me for restoration was all filmed back-to-back in a day or two, outside an aircraft hangar at Kemble Airfield.

He'd come to the workshop for the 'midway' pieces in each episode, of course. This was when he would turn up to deliver a vital part that he had dug up and 'complain' that I had done nothing, and I would explain what I had actually been doing. Like the handovers, we would fit several of those in a day. In the end, after my restoration was complete, I handed the cars back to Mike at the workshop before he took them out for the final test drive and sale. Again, we filmed as many of them in one day as was possible. So, during a three-month filming period, we might see Mike in the workshop for four or five days at most.

When Mike was around, we always had good banter but it wasn't until the later *Wheeler Dealers* series, after focus groups had recommended that I should be involved in the end test drives, that we could properly develop our on-screen rapport.

In 2007, having reduced the Cummfy Banana workload, we moved to a smaller workshop unit, on yet another farm, near Marlow. Discovery had not commissioned another series of *Wheeler Dealers* that year, and we decided that it was time to separate the two activities. As and when the show was commissioned again, we would look for a separate workshop studio to film in.

During our break from *Wheeler Dealers*, Mike and I shot a series called *Auto Trader* for Discovery Channel and then a series of *Pulling Power*, an ITV show that was produced by Mike's own production company, X2 Productions.

Then, just as we had got our new Cummfy Banana workshop sorted and were about to crack on with some new projects, we were given the notice that *WD* was up and running again. We needed a studio space and we needed it 'yesterday'.

I had noticed that the barn next door to mine was rarely used, so I tracked down the tenant and persuaded them to let us take over the rent and within a couple of weeks we were cleaning floors and painting walls ready for filming. The only trouble was the building itself wasn't entirely ideal; it was a metal barn with a corrugated roof and no insulation.

Inside, during the winter months it went down to -7°C and in the summer, temperatures reached 37°C. We had now extended the show and were filming almost continuously, so this meant I had to wear thermal underwear and several other layers under my T-shirts to keep warm during the winter and, for continuity, I would have to wear the same clothes in the heat of summer.

However, the most testing feature was probably the severely sloping floor. There was a two-foot drop from the back of the barn to the front so anything with wheels, including tool boxes, would quickly roll out of the workshop doors if not adequately chocked.

* * *

I may have had my *Dukes of Hazzard* moment in a Fiat 132 instead of a Dodge Charger, but the Charger is, unfortunately, another car I can claim to have crunched. It wasn't a spectacular General Lee-style smash but, just as those Chargers flew through the air in slow motion, I managed to crumple one at an excruciatingly slow speed.

Series 8 was the first time Mike imported cars from the States and an iconic muscle car had been on our wish list for some time. There is nothing quite like the sound of a big American V8 and the Charger has real presence and purpose even standing still.

The jobs on the car he bought included the power steering, the gear linkage, the rear suspension and the motor that operated the covers concealing the headlights. Under the dirt, the bodywork was a beautiful, gleaming black and really didn't need attention. Well, not up until I 'modified' it.

The Charger was on the ramp and we were midway through filming the jobs on the front end, with many of the parts now disconnected or dismantled. We had finished filming the scheduled job earlier than usual, at maybe five o'clock rather than eight or nine. Rather than call it a day, the new director, presumably keen to impress the producers, opted to film some 'pretties' – essentially, detailed shots of the car looking great. He wanted the car to be washed so we lowered it off the ramp and opened the roller shutter door to get it outside into the yard. I hopped in to steer it while Paul and the crew un-chocked the wheels and gave it a light push to get it moving.

The heavy Charger rolled forward into the yard and gathered speed. I tapped the brake pedal but nothing happened, I jammed my foot down hard, still nothing. Which was when I remembered that the brakes had been disconnected.

I was heading straight for a beautiful old barn so I yanked the steering wheel to the left, but with no battery in the car and without a running engine, there was no power steering. Unassisted, the steering was so heavy that the car wouldn't turn quickly enough, so I fumbled frantically for the handbrake lever between the seats. It wasn't there. Of course, like many American cars, the emergency brake is operated by a foot pedal.

I was just too late but it slowed the car down enough that the inevitable crunch did little more than lightly crumple the front right wing. Our prized Charger was now going to need a new wing and a paint job.

Finding a replacement wing was a headache. It could not be found anywhere in the UK and we had to source one from the States and have it shipped across the Atlantic at huge expense before we could start the repair work, none of which appeared in the finished show. We even had to film some of the ongoing work on the Charger, obscuring the damage, while we were waiting for the replacement wing. Oops.

In spring 2011, *Wheeler Dealers* had become Discovery Channel's biggest show, showing in dozens of countries around the world and the demand for the show meant that we were now filming full time, all year round. So, I thought we deserved a workplace with a break area, heating and level floors and I took the lease on an industrial building in Bracknell.

Attaboy rented part of the building for the studio workshop and production office, and the Cummfy Banana vehicles, which we were still renting out for events, and our office was in the other half of the building. Our MoT and Servicing centre, Grease Junkie Garage, was right next door. It was a great setup with all the facilities we could dream of, support when we needed from the garage - and we even had indoor plumbing!

We filmed three series of *Wheeler Dealers* in Bracknell until, in 2014, we were informed that we would be relocated to California for the start of filming on Series 12. The building had to go, along with everything in it and, eventually, the garage next door. We put the Cummfy Banana vehicles in storage and sold or scrapped a mountain of accumulated automotive treasures, which Imogen very uncharitably called junk. We even hired a professional eBay seller to manage all the postings and sales as we packed up our home and prepared for a whole new adventure.

10.

A WARM PROP

What's the last thing that you might expect when you are in the middle of testing a car for a TV show? A puncture? Nope, that could happen at any time. A breakdown? No, that's pretty much guaranteed. A numpty presenter running out of driving talent and piling into the back of the tracking car, from which the crew are filming you? No, the cameramen are always ready to pull their legs out of harm's way. You can expect pretty much any normal motoring mishap, working on the theory that 'anything that can go wrong probably will go wrong'. What you *might* not expect, however, is someone to try to land a plane on you.

We were filming for *Pulling Power*, an ITV show that I worked on with Mike and S.J. (Sarah-Jane Mee, whose proper job is hosting the *Sunrise* breakfast show on Sky News). We tested and reviewed cars on the show and had been scheduled to drive Britain's fastest electric car, which would have been extremely cool. Unfortunately, there was some sort of problem with it, so we never got to find out whether it really was cool or not.

We had an airfield booked and a camera crew ready to roll – everything we needed for a car test except an actual car. So the producers decided that we would instead do a piece on a car that had been built by one of the guys who worked for the production company. It was a kit car – an MEV Rocket – and it was, pretty much, finished and ready to go.

Rather like my Fugitive sand rail from years before, this car was basically a tubular structure holding together the suspension, wheels, seats and an engine – nothing much more. Essentially, it was a reasonably priced 'interpretation' of an Aerial Atom based around a 1.6-litre engine from a Ford Focus.

However, the seat position couldn't be adjusted, which made fitting the length of my arms and legs in behind the steering wheel almost impossible, so I took the seat out and replaced it with the foam cushion from a garden lounger. Once I had squeezed in, strapped on the seatbelt and held myself in place with my elbows clamped over the tubular frame, I got the car out on the tarmac and hooned around in front of the camera, chatting about what the car was and how it handled.

As usual with these things, we filmed all the 'talking' bits first and then the director wanted a series of shots where I would come haring past the camera, closer and faster each time. We had done all the drifting and power-sliding at one end of the airstrip and the cameraman now had a great shot down the length of the runway as I came hurtling towards him.

Camera operators need nerves of steel, as to get the most dramatic shots the driver needs to get as close to them and the camera as possible without actually looking towards the lens or indeed wiping them out. Rented camera equipment is usually well insured, and therefore if it is damaged or destroyed it is just a waste of valuable filming time before a substitute can be found. Directors of photography, however, are rather more valuable and frankly make much more of a mess.

On my final run I was going as fast as the car would allow, concentrating very hard on looking ahead with steely determination and not wiping out the cameraman when, as I flashed past him at around 100mph, I noticed something just above my horizon – a light aircraft coming in to land on the runway!

My mind froze for a moment; had the DOP just narrowly survived death by 'shot of the day' only to get flattened by a light aircraft? I couldn't do anything to warn him – he was still, no doubt, absorbed in capturing the dust settling in the light of a crepuscular ray, or something equally arty, and the deafening sound of the kit car's exhaust directly behind him was certain to drown out any hint of the impending doom.

A critical moment later I discovered that balancing and fine tuning the Rocket's brakes was one of the things still left on the car's 'to do' list, because when I stepped on them, they immediately locked up. Cadence braking did little to lessen my progress so I just jammed the steering fully to the left hoping to scrub off the speed sideways. The Rocket refused to change velocity (speed or direction) and I found myself careering towards the end of the runway, where an ominous grass embankment was looming large.

Fortunately, the plane's pilot realised that the runway wasn't clear and aborted at the last second. I have no idea where he or she eventually landed.

However, by then, I was rather more preoccupied with the danger of becoming airborne myself, if the Rocket shot up the embankment and took off just like its name suggested it would. Luckily, sliding up the bank, tearing long grass as it went, slowed the car and it stayed loosely in contact with the ground as it crested the rise and slithered down the other side, coming to rest in the bottom of a big ditch with a dull thud.

After the howling of the Rocket's engine, the buzz of the plane's engine, the squealing of the tyres and the thumping of the suspension had stilled, there was a blissful silence. I took a quick inventory of my fingers, hands, feet, limbs, teeth and everything else that can go missing in a car crash, but it all appeared to be where it should be.

Once everyone recovered from the surprise of what had just happened, and finished laughing at me of course, we recovered the car and found there was actually very little damage done. We packed up as if it was just another day filming. And despite a plane nearly landing on my head, it sort of was, really.

* * *

I'd never intended to build a career in TV. Especially considering how I felt about any kind of public performance. Given the crippling anxiety that had always descended on me when I was expected to perform in front of any kind of crowd, it is strange I feel quite comfortable in front of a TV camera. It didn't even occur to me that it might have been an issue. It was only when a friend of mine completely froze while being interviewed for a vox pop item that I even thought about camera fright. Even seeing it happen it didn't occur to me that it was the same affliction, just with a different trigger.

It might have been because I'd had good experiences in front of a camera before I noticed it was still a performance. After we had left school, Steve and I went back in the holidays and made a very stupid film full of in-jokes about one of our friends, called *The Legend of Toby Chiles*. It meant a lot of time in front of the camera in a very familiar place, mostly just talking to Steve and laughing a lot.

My first proper experience in front of a TV camera was just after the sofa had been featured in the *Sun*, when Sky News came to do a story. All I had to do was drive around a bit and answer some questions, everything was fine. Much later, when I started filming for *Wheeler Dealers*, Woody and Dan, the producers from Attaboy, told me to think of the camera as a person and to just talk to them. It certainly helped.

Filming live, on the other hand, brought back the anxiety in a flash. Damon Hill's team filmed our record-breaking antics at Donington for his video, *Damon Hill's Wild and Whacky Races*. As part of the launch activity, Damon drove the bed and bathroom around *The Big Breakfast* garden, tearing up their precious patch of grass. It was a fun piece, and even though it was all about Damon I was still stricken enough to leave my breakfast in their flowerbed.

Even before I went off to Belize I had racked up quite a portfolio of TV and media appearances with the Casual Lofa. That had, of course, been the purpose of it – getting paid for publicity to fund my trip. The sofa and I had appeared in features in local and national press, the *Sun* and the *Daily Mail*, several news and magazine TV shows, including *The Big Breakfast* on Channel 4. I had also done several interviews and photo shoots for magazines that were published while I was away, including *Custom Car Magazine*, which was particularly cool as I had been part of the custom car scene and a regular reader of the magazine.

So, at some point, the Casual Lofa caught the attention of the *Top Gear* team, who were planning a brand new venture, *Top Gear Live*. It was to be a high-octane motoring live arena show featuring car displays, burn-outs, doughnuts and stunts, all accompanied by the usual pithy banter of the presenters. And they wanted the Casual Lofa to join the cast for the event.

I had previously been featured in *Top Gear* magazine with the Furry Ugg, so they must have had my details and got in touch – and got through to Mum. I was deep in the jungle at the time though, and couldn't be reached. They must have been quite persistent as Mum eventually got hold of Dave, and Dave somehow managed to get hold of me, via that jungle satellite phone.

So I agreed to cut my South American adventure short, for a three-day paid gig in the Casual Lofa, and a date with Jeremy

Clarkson and Tiff Needell, who would both be taking the Casual Lofa out on the track.

The first words I ever said to Jeremy Clarkson, as he sauntered over at the start of rehearsals for his lesson in sofa driving, were: 'Alright, Short-Arse?' He is only six-foot-five, a good two inches shorter than me, and it seemed like the right way to start the conversation and set the right tone for the proceedings.

Jeremy loved the Casual Lofa and took to driving it straight away. I was impressed: the sofa's controls are radically different to those of any normal car and yet he just took it all in and was on top of driving it before his bum had warmed the leopard-print fur.

Tiff Needell, on the other hand, not so much. He's a racing driver through and through, a splendid chap, but incapable of doing anything but pushing a vehicle to its absolute limit. He simply does not possess a fear gland – or seemingly any mechanical sympathy.

For each performance, Tiff jumped onto the sofa and then slammed it into gear at full revs. The sofa would lurch into the arena to raucous applause. The crowd loved it, but the violent gear selection took its toll. The Casual Lofa was subjected to Tiff's punishment three times a day and by the end of the weekend, the auto box was slipping terribly. The upside, however, was that Tiff's harsh handling eventually led to us setting our first world record.

Dave Bickers and his team from Bickers Action were in charge of the stunts for the show. They normally provided special effects and action vehicles for big movies such as Bond so they made easy work of firing Tony Mason and his Metro through the air into a ball of fire at the end of each performance. Dave noticed what had happened to the Casual Lofa and offered me a 'one careful owner' Mini engine and auto box.

I drove over to their HQ in Suffolk and it turned out the 'careful owner' was in fact Eddie Izzard, when he played a baddie in the 1998 *The Avengers* movie. Bickers had removed the engine and gear box before hurling the Mini off a cliff for a scene in the movie. It was a more powerful, fuel-injected 1300cc motor, much better than my tired old 998cc. Dave and his son Paul let me have the engine and box for a very reasonable price, and so my mobile living room became a lot less sedate. But not, unfortunately, before we filmed a race for *The Most Outrageous Jeremy Clarkson Video in the World ... Ever!*

Following the Silverstone show, Clarkson had asked if the sofa could appear in a wacky racing video he was shooting at Thruxton. I was up against a motorised skip, a three-wheeled speedboat and a garden shed. We lined up on the grid, started our engines and waited for Jeremy to start the race.

I put the sofa into drive and went absolutely nowhere. I revved the engine a little more, then a lot more, and ever-so-gradually I left the start line. Once I had got going I did manage to start gaining on the rest of the racers but in the end I finished third behind the speedboat.

As soon as I got back to the workshop, I got to work installing the 1300cc engine and managed to get it working just in time for our triumphant race at Donington where I set my first Guinness World Record.

Once on *Top Gear*'s radar, I did quite a few other things both for their live shows and the TV show. I appeared as a boffin on one show in 2002, having created a 'Bond Car on a Budget' out of a Rover 820i. My Bond car had a tea tray fixed under the spoiler that popped up from the boot as a bullet-proof screen; a trap-door in the boot that showered the studio floor with coloured balls (like you'd find in a children's ball-pool play-pit); paintball guns hidden inside the front wings that fired out of the indicators when you pulled the furry dice hanging from the rear view mirror; and rocket launchers made from drainpipes stuck along the side of the body.

Clarkson provided us with the perfect target – a cardboard cut-out of his nemesis, the transport minister, John 'Two Jags' Prescott. Having shot him in the groin with our paint-balls, the rockets then took his head off to rapturous applause from the studio audience.

Of course, every self-respecting Bond car needs an ejector seat. I wasn't allowed to use explosives or compressed air in the studio, so I had six huge springs made that had to first be compressed. I made a special tool using a hand-operated trailer winch that would compress the springs one at a time. I had also made my own 'bomb releases' using roller blade bearings and some lengths of steel, with M8 bolts for hinge pins. As each spring mortar was armed, I locked it in place with a pin for safety.

Once all of the springs were armed and in position on the floor, the hollowed-out, front passenger seat could be placed back into position. A domestic brass doorbell simultaneously energised six Beetle starter solenoids which pulled on the bomb releases to set the springs free, firing the seat through the roof. There was a lot of energy stored in those springs – they were probably far more dangerous than just using a small explosive charge.

Instead of a roof that released and flew off, as on 007's Aston Martin, I cut out a large section of the roof and replaced it with foil, scored for easy breaking, and painted it in the car's body colour. The mannequin in the passenger seat was far heavier than I had anticipated, but I was sure he would punch through the foil.

It worked perfectly ... which means barely. When Richard Hammond and I retreated to a safe distance and pressed the doorbell on the end of a very long extension cable, the dummy shot through the roof and bounced off the back of the car. Its short flight seemed suitably crappy for a car that was supposed to have been built on a budget of £300.

* * *

Over the years, news of the crazy vehicles seemed to spread around the globe and TV crews from all over the world got in touch wanting to make a video. I was out one day, filming the Street Sleeper with a French TV crew. We had been doing some tracking shots on a duel carriageway outside Farnham and then the camera car pulled behind me to pick up some close-up shots of the back of the bed.

We got to a roundabout at the top of a hill and I could see the driver waving at me. I continued on, trying work out what his hand signals were the international sign for. A little further along the road, both the driver and the camera assistant were now waving and pointing at the back of the bed.

Everything seemed fine, and the bed was driving beautifully, but I figured something must be bothering them so I looked for a place to pull over. Further up the road on my right was a petrol station, so I indicated and pulled in. The garage was very busy so I parked up next to the tyre inflator and water dispenser.

By this time the French chaps seemed beside themselves with excitement so I got out and had a look at the back of the car. The lower drapes were engulfed in flames!

I lunged for the water hose, only a few metres away, and sprayed the back of the car. For a worrying while, the water only spread the flames further around the back of the bed, but eventually the fire went out, just in time to stop a member of the garage's staff from dousing me and the back of the bed in foam from a pair of fire extinguishers.

On closer inspection, through the smouldering, steaming fabric, it seems a small oil leak had been gently soaking the drapes and then the heat from the exhaust had been the source of ignition. It meant Mum would have to run up a new valance but it made for a great piece for French television, so I was 'appy.

Around the same time I shot a music video for Mint Royale's single 'Take It Easy'. I was dressed in a particularly stylish lounge lizard's suit and all I had to do was drive the Casual Lofa around

London for a couple of days. It was brilliant fun and came with an unexpected bonus. A few months later I drove the sofa down to Cannes in the South of France for a holiday and one night I parked it outside the campsite's bar. The revellers were impressed and offered to buy round after round of drinks. As the evening wore on, the novelty wore off but just as the free drinks started to dry up, the 'Take It Easy' video appeared on the bar's big screen and with drunken cheers and applause, the free alcohol was flowing again.

The plethora of random news items soon brought me to the attention of other telly people and I was offered the chance to try my hand as a TV presenter. In 1999, I was asked to co-present a pilot for a show intended to be an upbeat, irreverent, rock 'n' roll version of *Tomorrow's World,* together with Gail Porter and Vinnie Jones. Its working title was *Boys Only* and they had us presenting a variety of features that involved real science illustrated with various stunts.

I remember Gail getting covered in fake blood during a medical piece and Vinnie parking a tank on top of a car. All the producer seemed to want me to do was to drive around in the sofa. The show never got beyond the pilot but it gave me a taste of what was to come.

A couple of years later, I worked with Gail again for an episode of *Pulling Power*. Two of the regular Cummfy crew, Jonny and Rachel, helped me to drag pretty much all of our street-legal vehicles up to London for a gloriously sunny shoot. The item ended up being a brilliant ten-minute advert for the creations of Cummfy Banana.

By now I had already racked up my '15 minutes of fame' and friends and family were phoning me increasingly regularly to point out on what channel or in which magazine they had seen me. I was starting to see how exposure like that brought with it some kind of infamy and could easily feed a desire for recognition.

I have worked with celebrities who crave the adoration of their public and are desperate for attention. It is not a healthy pursuit and wreaks havoc with your sanity and peace of mind. Thankfully, I have always found that should your ego get over-inflated, there is always someone close to you, or just around the corner, who can bring you back down to earth with a bump.

I was in south London one day, at a petrol station filling up the sofa and, as I approached the kiosk to pay, a woman walked up to me and asked for my autograph. I was completely flabbergasted, why did she want my autograph? Maybe she thought I was someone else?

She thrust a pen and paper in front of me and I then started to panic about what to write; I only had one signature, the one I use on cheques and official forms. It didn't occur to me to write anyone else's name, I suppose I could have made anything up. I went ahead and signed a squiggly version of my name.

As I looked up, handed the woman back her scrap of paper and thanked her, I noticed there was now a queue of two more people. What was going on? I signed the second autograph with a little more confidence, and then the third with a veritable flurry. Receiving her priceless trophy, the woman looked down at my scrawl, looked up at me and stared for a moment, eventually uttering the priceless words: 'Who are you?'

Closer to home, one of the producers of *Wheeler Dealers*, Woody, would often refer to me as 'the warm prop', presumably so the rest of the crew wouldn't confuse me for a light stand or some other bit of set dressing. In the States we were referred to as 'talent', seemingly with the word in quotes – as, after all, we were mostly just standing and talking, walking and talking, working and talking or even driving and talking. Suffice it to say, I always have plenty of help to stay grounded.

* * *

My 'real' job at Cummfy Banana required me to conjure up ways to repurpose the workings of automobiles to power all kinds of un-car-like objects. Cathy Rogers' *Scrapheap Challenge*, on Channel 4, had two competing teams of engineers do much the same, creating all kinds of contraptions out of car parts and other junk left in a breakers yard. I was honoured to be a guest and judge on a few of the episodes over the years.

In around 2001 Sonia Beldom devised a show for BBC2 called *Panic Mechanics*, which had a similar theme. Two teams would be repurposing old cars for new challenges but with a more aesthetic approach and a twist at the end: the vehicle that lost the challenge would be crushed.

Each team would have a qualified team leader and the rest of the team would be made up of members of the public. When they were looking for suitable engineers and car builders for a pilot, my name came up everywhere they looked.

So, in early 2001, I found myself in one of two adjacent workshops in the pit lane of Donington circuit, shooting another pilot. In the other workshop was legendary car customiser Andy Saunders, and we were supported by the great Sean Cunningham, who usually helped Diarmuid Gavin with engineering on garden shows like *Ground Force*.

The show host was Trevor Nelson and he presented us with two old Rover Metros to convert into, well, 'something interesting', fit for a challenge that I am not sure we even knew at the outset.

I dragged Paul along to give me a hand as usual, and we decided to create a kind of shortened, Monster Metro with huge balloon tyres, an open top and a roll cage. Our opponents next door had opted to graft a fibreglass, Audi TT-looking kit car body onto their Metro and then modify things from there.

It was intense work but loads of fun with great camaraderie. Andy helped us out as we neared our 48-hour deadline and Sean amazingly managed to source some custom-made wheel adaptors for our huge wheels overnight.

To cover our welding sins, Paul and I painted the whole car in satin black paint – which would become a theme in future *WD* episodes – and added a few yellow stripes, which led to the car's name of Tiger. Because of the huge diameter of our new tyres we didn't fare so well on the drag race challenge, but in the mud of the auto-cross course we triumphed.

The pilot was a success and, in the autumn of 2001, we shot a seven-episode series for BBC2 at Pebble Mill Studios in Birmingham. I was now the show's resident designer and team captain and was up against three opponents over the series: kustom car creator Andy Saunders, engineer Ranen Rudra and hot rod builder Paul Burnham.

Each week, our two teams were given an identical donor vehicle which they had to re-purpose to tackle a set challenge, with the winner being decided by some kind of race. The 'panic' bit of the title was because we were only given 48 hours to transform our vehicles.

The teams could win extra parts or budget for their build by setting the fastest time negotiating a course in Daisy, a hulking great rock-crawler style buggy built by the rightly renowned engineer Ralph Hosier. The controls were shared between the two 'drivers' so communication was key.

In one of the episodes, we were given a diesel Ford Escort van and told to turn it into a gravity racer. A gravity racer is simply a sophisticated soapbox cart, but real racers are slick and streamlined and capable of over 80mph. Somehow, I didn't see us getting anywhere near that speed, rolling down the hill at Donington, but speed wasn't our aim.

Our challenge was to roll our modified vans down a long hill and then up the other side. Whoever got furthest up the hill would win. Relying on 'pub physics', I figured if we left the deadweight of the engine in and carried an extra tonne of water in the car on the downhill run, we would build up lots of momentum and then, by ditching that water just before we hit the uphill section, that momentum would carry the newly lightweight vehicle further.

I decided to go with a nautical theme, hoping to emulate the lines of a motor cruiser. We made a pointed rail like you would have on the bow of a boat and we cut down the back (OK, the stern) to fit our huge plexiglass box of disposable water. Andy's design was based around using a huge 747 wheel as a kind of energy storing fly-wheel, but in the end, my pub physics won the day.

Everybody put a lot of effort and ingenuity into those cars and I always found it upsetting that *Panic Mechanics*' formula demanded that, while the winner took the plaudits, the losers had the trauma of watching their cars being crushed!

On that particular show, I had marked out the new 'cruiser' profile along the side of the van with masking tape and was cutting along it with an angle grinder. The cameraman was filming the sparks flying as I made my way towards the lens. Having noticed him there, I finished the cut, turned to him and briefly explained what I was doing. The camera guy was really excited that the piece to camera had finished off the shot nicely. It just seemed very natural to me to explain what I was doing.

Working and chatting are two things I have always found very easy. Even when I was little, my granny couldn't believe how much I talked when I was sat at the foot of her bed, building all kinds of random things out of Lego, without any instructions, and chatting constantly not only about what I was doing but about everything else besides.

Although, that's not to say that I don't sometimes struggle to explain a complex process or principle in simple terms, in just the right number of words, while also showing the camera exactly the right detail of a part or assembly. It can take me ages to get technical pieces to camera right!

I have now forgotten my all-time record of takes for a single piece to camera. It was definitely more than 40 (but hopefully less than 100). Thankfully, we rarely used a clapperboard with the number of the take chalked on it, so only the editors really know for sure. That, of course, is another thing that soothes performance anxiety on camera. You know that, within reason, you can run through the piece again and again until you get it just right.

All the builds on *Panic Mechanics* were great fun; we turned hearses into dune buggies and raced them on a private beach in Devon. We turned Reliant Robins into drag racers – the competition got a turbo, we bolted two engines end to end and added a sniff of nitrous oxide. We also modified Minis to transport all of the members of a brass band, with their instruments, around a course.

Our Mini was hopelessly rusty, so after the first day of filming, Paul and I went to get something to eat and then returned to the studio at Pebble Mill. We managed to persuade the security guard that we had to 'just pop in to collect something we've forgotten' and then worked through the night to repair all of the rot so we could carry on our conversion on camera the next day.

My cunning plan was to split the rear of the Mini along its centreline, hinging the two back halves outwards from the door hinges and welding in a new floor to make a wide loading space in the back. The resulting 'arrow'-shaped car would then have more room for the band members and they could jump in and out more easily. It did work very well but Run BMC, as we christened it, lost to Paul Burnham's super long Mini limo. Shame.

My first ever electric classic car conversion also featured on *Panic Mechanics*. Well, kind of. As part of the pre-production, we had come up with the idea of racing FX4 taxis across a stretch of water, with one that could go across the surface and the other driving underwater. Health and safety won out in the end so we needed another idea.

As a kid, I had often been taken to see the Royal Tournament, where the Royal Navy field gun competition would see two teams of sailors take apart a field gun, race the parts over a course, through walls and over a chasm, and then reassemble it, firing a blank to finish. It was great entertainment, so my suggestion was to do something similar with a car.

My idea was for each team of four to drive in their car up to the front door of a house, take the car apart, carry the bits through the house, out the back, and then rebuild it for a short race to the finish. Beetles seemed to be the obvious choice for the donor cars as they were bolted together by hand in the factory.

I opted to replace the heavy body shell with a kind of flat-pack corrugated plastic design. Ranen went down the route of a high-tech dome tent design with lossless connectors for the brakes and fuel, and big over-centre connectors to hold the various parts of the floorpan together.

Paul and I spent most of the first day dismantling our Bug into smaller parts and trying out how heavy each lump would be to lift. Fewer parts taken off meant fewer trips through the house but made each lump heavier to lift. It suddenly occurred to me that we might not actually need the engine. What if we just used the starter motor, like you can if your car conks out at a junction? You put your car in first gear, turn the starter and then you can steer your car out of harm's way.

I realised we didn't even need a clutch pedal, we could just select a gear and leave it. If we added some batteries, we could probably get enough range too. We created a box for the car batteries, which was to double as a seat, and then used stretcher poles to carry it around. Paul made a bracket to hold the flywheel onto the first motion shaft coming out of the gearbox and we added a button somewhere to operate the 'throttle'.

For the body, we created a pair of Beetle-shaped 2D 'cartoon' sides out of thin, corrugated plastic boards and then made the roof and wings by stapling loads of pieces together to make the curves, rather like the shell of an armadillo.

On the day of the race it all worked a treat, we woue-woue-woued our way up to the house, dismantled our cartoon Beetle in a jot, struggled our way through the house and set to rebuilding our car in the back garden. Ranen's team had won two marines to help carry their parts, but Ranen was then only allowed to shout directions rather than take part in the challenge.

Their extra muscle definitely helped them reduce the advantage we had gained from having less parts to lug, but all that was lost when trying to reassemble their 'cat's-cradle' of poles and fabric. So, my first electric classic was an absolute triumph. It only took me 15 years to get around to tackling my next – the electric Maserati Biturbo for Series 13 of *Wheeler Dealers*.

Panic Mechanics was broadcast in early 2002, drawing a very good audience by BBC2 standards of around three million. Despite this, it was cancelled after one series, even though US channel TLC had bought the first series and loved it so much they wanted to pay for half of the next series.

That was a shame, but we weren't the only BBC motoring show to bite the dust at that time. *Top Gear* was also pulled, which led to two of its producers, Dan Allum and Michael Wood, setting up their production company, Attaboy TV, and developing the concept for *Wheeler Dealers*.

TLC wanted the next series of *Panic Mechanics* so much they ended up creating their own version – *Monster Garage*. This time it was just one team of expert builders repurposing each vehicle. I joined a team to convert a milk float into a drag racer. We gave it a roof-chop to reduce the drag from its frontal area and replaced its tired electric motor with a big V8. It looked great, and went very well.

Perhaps inevitably, many years later, I broke a Guinness World Record for fastest milk float with help from Dick Bennetts' West Surrey Racing team and their BTCC driver Tom Onslow-Cole. The challenge was set by eBay and we had to buy everything we needed for the build through their site. We ended up with an old Cabac milk float, a Rover V8 from an old ambulance, the suspension and brakes from a Jaguar XKR and a load of tuning parts. It was quite quick and handled well and we made a fun gymkhana video with Tom, racing around some derelict army barracks in Aldershot.

* * *

Panic Mechanics was a fun and interesting experience, and I thought making TV shows seemed fun, but Cummfy Banana was still my main focus and my bread and butter. So, when Dan and Michael first asked me to do a screen test for a new show they were working on, I was curious but way too busy.

They wanted me to come to a barn on the other side of London and I simply didn't have time – I was working my pants off to build the driving office with the deadline looming. In fact, had *Panic Mechanics* ever wanted to film an actual mechanic in a real panic, they only had to pay a visit to the Cummfy workshop on any given day!

Luckily, fate took a hand. The desk top I had designed for the Hot Desk was made up of a number of pieces, all of which had curved profiles and then radiused edges top and bottom. So I needed a company that could accurately cut out the shapes from my CAD drawings on their CNC router. As the office car would spend much of its time out in the open, I also needed to use marine grade wood and for it to be protected with layers of marine varnish.

The specialist company I found were based in a business park on the other side of London ... coincidentally just a few minutes from where Attaboy wanted me to do their screen test. I had emailed the designs through weeks before but, when I turned up to collect my desk top, it wasn't ready. I would have to wait a good few hours before I could take it away. This was not good, I was on a very tight schedule, but there was little point trying to get back to the workshop and sitting in traffic for hours in an attempt to do a little extra work. Then I remembered the screen test.

So I figured, OK, why not? Attaboy had asked me to bring along my favourite tool, so I rummaged around in the boot of my car and found my new welding mask and set off to kill two birds with one stone.

For the screen test, Dan and Michael had set up a Land Rover in a workshop and asked me to change a wheel, while explaining what I was doing. I talked through the whole sequence and, when I came to replace the wheel, I explained my usual trick of sitting on the floor facing the wheel arch and supporting the wheel on both feet – this makes it much easier to adjust the height and position of the wheel when aligning it with the hub, and it leaves your hands free to wind in the nuts or bolts.

They then asked me to do a piece to camera about my favourite tool. I explained how my welding mask had an LCD auto-darkening feature which was triggered, in milliseconds, by the bright light of the welding arc. Such a simple idea but it's one of those tools that transforms your productivity. Dan and Michael said they would be in touch and I toddled off to find out how my office's desk top was doing.

A few weeks later, they called asking if they could pop down to my workshop in Odiham to film a short piece with Mike and myself, to see how we would get on together.

It was immediately obvious that there was great on-screen chemistry between Mike and me. It also created interest that we weren't what people expected in the roles that we were playing. The car dealer is often seen as the slightly posh, well-educated entrepreneur, whereas the mechanic is usually the cheeky chappy with a barrow-boy accent, but with us, those roles were reversed.

It was clear that this juxtaposition was going to be entertaining as we discussed my old Range Rover for the camera. Mike was pretending he had bought the car and brought it to me, just like he would do on the show itself. He was at a disadvantage, though, as I was easily able to pick fault with his purchase – I was already on intimate terms with all of that car's many problems!

We started filming the first series of *Wheeler Dealers* in May 2003 and even our height difference added to our 'chalk-and-cheese' partnership. On the first day of filming the show we were in a hangar on Kemble airfield, discussing the Saab Mike had just bought. Dan or Woody offered up a pallet for Mike to stand on, to help even up the height difference on camera. Mike made it pretty clear that was never going to happen, and rightly so. From then on we just made sure I always stood the furthest away from camera and wore baggy, layered clothing.

11.

MAKING A SPLASH

If anybody asks me where my obsession with amphibious vehicles comes from, I blame it on the Duck. The DUKW were amphibious six-wheel-drive trucks with a watertight hull built by General Motors during the Second World War to ferry troops and supplies from ship to shore. They were used on the beaches at D-Day and in the island campaigns in the Pacific. The letters are a fairly meaningless manufacturer's project designation, but thcy lcd to thc vchicles being affectionately known as 'Ducks'.

I first encountered a Duck on holiday in Cornwall with my family. We had wonderful family holidays staying with my grand-parents and aunts in Mousehole when Clare and I were kids. For a treat we would be taken to St Michael's Mount. If you've never been, this is a real-life castle, sitting on its own island off Marazion Beach, built of solid granite with turrets and towers, battlements and cannons and a 'secret' stone causeway that is only revealed when the tide goes out. Real-life battles had been fought here, by knights wielding swords to take control of the castle. Legend has it that it was built by a giant back in the day when giants walked the earth, terrorising the locals and shaping our landscape by pelting rocks at each other across the English Channel ... How could a small boy not be enthralled?

The best part, though, was if I could just dawdle long enough and hide away until the tide came in, the causeway would flood and the only way off the island was by this massive amphibious craft which I was utterly fascinated by. The Duck in question was at least 30 years old at that time, it was noisy and smelly and slow, but I thought it was amazing that it could go from land to water.

Cornwall has always been a magical place to me – a land of legends and knights, pirates and smugglers and lots of secret places to explore. As a wide-eyed boy, it provided a perfect backdrop for my curiosity and rampant imagination to be nurtured and grow. I usually brought a friend to stay for the holidays and, growing up in the days before health and safety and helicopter parenting, we were given lots of freedom to roam. We'd explore caves along the rugged coastline, retrace the steps of Arthur and his knights at the ruins of Tintagel and spend endless days playing on the beach and in the sea.

Being near the sea played a big part in my young life. It fostered a passion for getting out onto, or into, the water, and not always in the most orderly fashion. When I was older I would get into windsurfing and power kiting, but when I was little, I would jump in anything that would float.

I remember going for a walk with my sister, our friends and our mothers when I was probably only about nine or ten. We were meandering along a river bank, splashing in and out of the water, when we spotted an old bath stranded on a mud or gravel bank. It must have been ripped out of a house that was being renovated, but goodness knows how it had ended up in the river. I suppose that some kids just like us might have launched it into the water further upstream and the old bath had become a boat. Well, clearly it had floated away from them, so now it was *our* boat. It was easy enough to get to, through water that wasn't even up to our knees. The bath taps were long gone and the plughole was bunged up with mud and grit.

My friend, Charles, and I started bailing water out but got told in no uncertain terms that 'the girls should have a go first'. So we hopped out and Clare and her friend Alice clambered in. They scooped out a bit more water, started to wobble when the thing finally floated free and almost immediately capsized, deeming our proud vessel a lost cause. So Charles and I were back in charge and we edged our boat out to a spot where the water was deeper, bailed out as much as we could, and set sail.

Perhaps because we were heavier than the girls the thing was actually now quite stable and, to our delight, we started to pick up speed in the current, heading downriver. At some point, we took a look at where we were going and we could see that the sea really wasn't far away. Then we took a look at where we'd been ... and the family was fast fading into the distance.

We knew that we weren't equipped as a seagoing vessel, so we had to stop this now rather swiftly moving bath/boat. It had no anchor and no brakes so we tried to capsize it, but by now it had become incredibly stable and it just wouldn't roll over. Looking back, I suppose we should have been panicking, but we were treating it all as a great laugh.

Eventually, by throwing ourselves from side to side in fits of giggles, we managed to tip the bath over and abandon ship. Luckily, we weren't too far from shore. We were completely soaked but that happened most days during our Cornish summers and we didn't even get a scolding. I guess our mums were quite glad to have us back.

* * *

After the DUKW, the next amphibious detail that captured my imagination was the hovercraft. There was something magical about it. A craft that could go from dry land out onto the water without even pausing for breath, and it is fast on both – what's not to love? As a child, I watched the hovercraft many times from the shore and was utterly captivated.

It is a great British invention that was never properly carried forward and developed. They're super-fast and can operate in places other craft can't access, but the problem is that they're power-hungry and very noisy – which means they're expensive to run and not suitable for densely populated areas, and therefore not always viable for commercial use. But even so, what an awesome feat of engineering – to bridge the gap between two elements using a third! Using air to travel across earth and water, well it's almost poetic, and I simply love the idea of straddling two physical realms in defiance of conventional ideas of what is possible.

Design-wise, the hovercraft does still have its limitations. For example, I know that its inventor, Sir Christopher Cockerell, never intended the fragile skirt that holds the air cushion to be a permanent solution. If the skirt is damaged or torn, the air cushion loses pressure and the hovercraft ceases to hover. I believe Sir Christopher wrote a paper on how it could be done using an 'air skirt' (much like the ones we now use to dry our hands in posh public conveniences). Unfortunately the technology was not available at the time, but, more than half a century since he designed his first hovercraft, and with the advances in technology, materials and manufacturing techniques, we must surely be able to improve on the original. I would love to get my hands on that original paper and find a way to realise Sir Christopher's vision. I might even have a go one day.

I even heard a rumour that there was a script for a follow-up to *The Italian Job*. Apparently, instead of using Minis as getaway vehicles, they wanted to come up with an alternative equally quintessentially British vehicle. The hovercraft was top of the list. Good choice!

One very special amphibious vehicle that I was lucky enough to take a ride in was arguably the most famous car on the planet – Chitty Chitty Bang Bang. Blink and you might miss it, it was such a brief moment, but in the movie Chitty did transform into a hovercraft to escape the tide and the Vulgarians. I loved that car when I was growing up.

When I was presenting on a series called *Classic Car Club* on Discovery, I suggested that we should do a piece on a movie star car. At the time, I thought the producers would go for Nautilus, Captain Nemo's car from *The League of Extraordinary Gentlemen* – a 22-foot, highly ornate, six-wheeled work of art. The movie was fairly current and the car was enormously exotic, based on the chassis of a Land Rover fire truck. To my delight, however, they went for Chitty.

I thus got to have a go in 'Gen II' the roadgoing car that neither flew nor turned into a hovercraft (the one that did that in the film was a shell on a speedboat hull and was destroyed after filming) and be driven round by the guy who then owned it, Pierre Picton, who came dressed as Caractacus Potts. Pierre was a circus clown and had driven Chitty in the movie and producer Cubby Broccoli gave him the car as part-payment for his services. He had used the car in his act ever since.

I was relegated to the passenger seat as the production company didn't want to cough up the funds to insure me to drive, but it was still a real thrill to be in Chitty Chitty Bang Bang. Gen II was later bought by the director Sir Peter Jackson, who had her shipped to New Zealand, where I am sure he will keep her in pristine condition.

We have a number of lovely Chitty replicas here in the UK. You can go and see one at the National Motor Museum at

Beulieu. It's also been a real treat for me to drive Chris Evans' Chitty replica at various events over the years.

It was also on *Classic Car Club* where I finally got to drive, or should I say pilot, my first real amphibious car – a beautiful, bright yellow 1960s Amphicar. Amphicars were manufactured in Germany from 1961 to 1967. Just under 4,000 were produced, most of which went to USA. There are around 500 thought to still be in regular use – 7 in the UK and about 80 in the rest of Europe.

It was an incredible experience, pure exhilaration. We had a perfect summer's day for filming and I was incredibly excited sitting in the Amphicar on a grassy bank waiting for the signal from the director. There was a film crew bobbing around on the lake and when they were ready, I got the signal to drive forward down a fairly steep embankment straight into the water.

It doesn't feel right. You know that you shouldn't be doing it. This is a *car* and you've just been driving it on the road. Surely you can't just hurl it into the water?! Of course, the amazing thing is ... you can.

The slope I went down made it feel as if I was going straight to the bottom of the lake but, as I entered the water, the Amphicar's nose bobbed up and we burbled off, the little Triumph Herald engine buzzing away to turn the twin propellers in the tail. I was over the moon. It was just the best filming day ever and I knew that I had to do that again. That car was a joy.

The next time I had a go with an Amphicar was when we gave one the *Wheeler Dealers* treatment. That car was *not* such a joy.

We knew that there were some bodywork issues with the car when it was bought in Florida but these cars are in great demand and, therefore, they don't come cheap. Finding one at a price that would suit our budget meant settling for one that had a few problems. That, after all, was what the show was all about. Mike and I had the usual banter when we filmed the handover back in the UK. I joked about having a 'sinking feeling', but at that point I had no idea what a state the car was really in.

The poor old Amphicar had bodged repairs everywhere. Bubbling paint on the rear three-quarters was a bad sign, but when I started stripping it back I was amazed at just how much filler had been used and it started to dawn on me that I was in for a bodywork nightmare.

To cover up a crap repair job, the entire panel had been skimmed with filler, like a plasterer skim-coats a wall. It was the same all over the lower half of the car. There was fibreglass, expanding foam and filler everywhere. In some places the filler was up to an inch thick, layer upon layer.

I couldn't believe that Mike had merrily taken this thing into the water in Florida. The filler is heavy and must have affected the handling; it does not float, so must have affected the buoyancy, and it is porous, so must have been letting in water. Mike is lucky that he didn't finish that test drive up to his armpits in alligators! At this stage, though, we were committed to the job, so all I could do was roll up my sleeves, pull on my orange gloves and crack on.

I knew that we were going to have to strip the car back to bare metal (where we could find any) to tackle the rot. This would normally be a very boring process that would take days and make for dull viewing on TV. What we needed was something that would look a bit more interesting for the camera and get the job done quicker, like *blasting* the paint and filler away. To make sure we didn't blow the whole car to bits, I researched various gentle blasting methods, like CO_2, or dry-ice blasting, which is used to clean fuel tanks and jet aircraft. It would have looked spectacular on screen, with clouds of the stuff billowing all over the place, but it was just a little too expensive and highly specialised for our project.

In the end, we settled for soda-blasting, having found a team who could bring their kit to us so that I could be filmed tackling the job outside the workshop. Actually, and rather cleverly as it turned out, outside somebody else's workshop! It looked just as great on screen and got the job done but it was a really messy process, definitely not something to be done indoors or on your drive at home.

It was worth it, though. The soda blaster stripped away the paint and filler, leaving the Amphicar's bare metal bodywork clean and smooth. It also created huge clouds of soda, filler and paint, turning what had been a decent day into what looked like a Dickensian London pea-souper. There was a carpet of dust lying on the ground all around and a layer of the stuff over everything else in the vicinity, including other people's parked cars. Stripping the Amphicar took hours of work but the clean-up job afterwards took even longer.

When we came to film the Amphicar test-drive (well, maybe test-cruise), I had wanted to do it on the Thames near Marlow where they have an amazing nineteenth-century suspension bridge. Designed by William Tierney Clark, it was his prototype for the bridge that he designed to span the Danube between Buda and Pest. The producers, however, wanted to film further down-river so that we could have picturesque shots of the Amphicar with Windsor Castle in the background. That seemed fair enough – we had a pretty good day for filming and we *did* get picturesque shots with Windsor Castle in the background.

However, it wasn't exactly plain sailing. Recent rain and flooding meant that the river was running high and fast at about seven knots. The Amphicar, even with its lovely Triumph Herald engine giving it the beans, could probably manage about 6.5 knots. That meant we would be able to make good progress going with the flow, but against the flow it would be a bit like swimming up a waterfall.

We turned up at the crack of dawn on the day of the shoot to make the most of the forecast early sunshine, for as long as it might last.

Before we were allowed on the water, we had to go through a safety briefing by the 'Admiral' of the rescue crew who were there to look after us that day. We sat down in a riverside briefing room, furnished with enormous bacon sandwiches and plenty of sauce. Delicious!

Just as we started tucking into our sarnies, a series of truly gruesome pictures of gnarled and gashed corpses with horrifically torn flesh, and heads and limbs missing, were flashed up on a screen. The admiral barked out a warning to keep our limbs, and presumably heads, tucked in to avoid the propellers of passing boats slashing us to pieces, as had happened to the poor devils in the photos. Well, that certainly put us all off our breakfasts.

The next most important thing that they told us was that, if we were in trouble while still on board our vessel, we should wave both arms in the air above our heads. We would have a walkie-talkie, so we could actually just tell them we were in trouble, but the manic waving would apparently let other boats on the river know, too. (In the event, though, there were no other boats crazy or stupid enough to be on the river in those conditions.)

Eventually we were ready to get the car in the water. When you're launching a ship, it's traditional to smash a bottle of Champagne on the bow, but after all the effort we had put into that bodywork, the last thing we wanted was to put a dent in the car. Instead, Mike opened the bottle and sprayed it over the bonnet of the car and we were ready to launch.

After I'd finished with the bodywork on the Amphicar, it had been sent off to be painted. This was the first time I'd seen it since and it looked pretty good, but to save time they hadn't put back the fender strips that run along the side of the car. Instead, to protect the paint, they had stuck on some strange sort of stone-chip film that really wasn't going to do much good at all if we were hit by a floating log on the water.

Neither would the film afford much protection to the precious bodywork when we came down the slip into the water and were hit by the current. The danger was that the nose of the car would immediately be swept sideways and the flank of the car might be bashed against the side of the slip, which would have been disastrous. To avoid this, we deployed a series of white 'bulb' fenders over the side of the car whenever we were going up or down the slip. As it happened, though, keeping the Amphicar's paint pristine turned out to be the least of our problems.

Once we were out on the river, everything was going swimmingly, so to speak, for a while. Our biggest problem seemed to be avoiding the various species of wildfowl on the water while the camera crew filmed us from the Admiral's gin palace motor cruiser. It was all very jolly. Until we turned to go upstream ...

The camera boat was racing ahead, with us at full pelt trying to beat the current and keep Windsor Castle in the background – a really cool shot. Then disaster stuck. My heart sank (fortunately, the only thing that did) when there suddenly was a clattering, clunking noise and I caught the unmistakable whiff of gearbox oil. Something had broken.

I switched everything off and Mike spoke the fateful words into his walkie-talkie, 'We've lost drive ... we're drifting ...' We waved our arms above our heads like we had been told and the rescue team sprang into some form of action. There were two boats – the Admiral's gin palace and a semi-inflatable, bright orange rescue boat. We had a gleaming white tow rope already fastened to the Amphicar, which was just as well because the tow point was underwater, so somebody would have had to get very wet to tie a rope to it.

Now, while I swim like a fish, a captain must never leave his ship, and Mike, I suspect, swims like a breeze block and certainly wouldn't have wanted to take a dip in those swan-infested waters. The rescue boat, however, had no suitable point to which we could attach the rope and not even enough power to tow the Amphicar with us in it anyway. I guess their job was to fish us out if we ended up in the drink.

Thankfully, the rescue boat still managed to edge us sideways towards the riverbank, where our superstar *Wheeler Dealers* mechanics Paul and Phil were waiting, ready to give us a hand. Phil caught the line and hauled us to safety while Paul went to get his infamous yellow Sprinter van to tow us back up the slip. The second we came alongside the bank, Mike was out of the car with astonishingly gazelle-like agility, I barely felt his weight as he stepped on my head to get across me and onto dry land.

And the problem? It turned out that during the test drive in Florida, the Amphicar had indeed taken on water, with some seeping into the gearbox through a breather hole, where it had been sitting for months during shipping and storing before it got to us at the workshop. This had made a small bearing in the gearbox go rusty and so it shredded when put under heavy load.

It was a pain and a disappointment on the day, but it was easily fixed with a replacement part before the car was sold. We never got the chance to take it back on the water for the camera again, which was a shame, because by the time we had finished with it, that 50-year-old car was running beautifully and looking as good and as seaworthy as the day it left the factory in Germany.

* * *

We may have saved the Amphicar from turning into an Amphisub, but the idea of a car that you can drive into the water and then under water is simply irresistible. Who could forget Roger Moore's Lotus Esprit in *The Spy Who Loved Me*? That was one cool car, if more than slightly implausible. The one that ended up in the water was actually a mock-up – a sealed, waterproofed empty shell. The one fired off the pier with an air ram was also an empty shell, but with wheels and everything still on it. So as it leaves the pier, you can see the car's running gear underneath, but as soon as it's in the water, it has a smooth, sealed hull on the underside.

The Esprit that was filmed for the underwater scenes had two scuba divers inside working the controls. It could swoosh around under the water using electric motors, but it only went forwards and it wasn't a car inside, it was just a fibreglass shell that filled up with water.

The original James Bond movie car was based on six reject fibreglass Lotus shells that were tarted up to make them look the part. It was all properly low-tech, as movie special effects often were in those days, but looked convincing enough on screen, and it must have inspired thousands of budding engineers to push beyond the plausible ... including me.

Elon Musk owns that Esprit now and has said that he'd like to turn it into a proper submersible car. The great news is that technology has moved on quite a bit since the Bond car was built in 1976, so it's no longer an implausible aspiration. It may not come as a surprise that I have lots of ideas about how it could be done. As I always say about everything, maybe I'll give it a go one day ...

Top Gear had a go at building a James Bond sub in their gung-ho way, but they used an Excel, not an Esprit. It's not nearly such a sexy body shape and the vessel they came up with wasn't really a submarine anyway. It bobbed about just below the surface, using pressurised air to keep the water out of the cab, but it wouldn't have survived going any deeper or for any length of time.

However, I do still think it would be possible to create a Lotus sub. I think the trick would be to start off with an original, period car, 3D-scan it and use that data to design and create the new body using modern composites that are much stronger. The primary focus would need to be on designing a submarine, then you could worry about making it work on the road.

I suppose, ideally, it would be great to be able to operate down to the recreational diving limit of 40 metres, but that would mean the hull would have to withstand five atmospheres of pressure. Quite a big ask. There is plenty of sophisticated scuba equipment available nowadays that could be used to operate the car in submarine mode: mini-thrusters and servos rather than frogmen and broom-handles.

It would have to be properly amphibious though; fully able to drive in and out of the water under its own power as well as handle equally well on the road. I'd definitely make it electric, and four-wheel drive would be essential so it could slip effortlessly out of the water and onto the beach; it wouldn't be very 'Bond' to be floundering at the water's edge.

* * *

In the meantime, however, I've got my hands full with amphibians that stay on the surface. For the past couple of years, I've been privileged to work on a project with a fantastic company called Gibbs Amphibians. Based in Nuneaton, they make high-tech, high-speed amphibious vehicles that, as you would expect, perform as well on the water as they do on land.

Gibbs first built an amphibious car called the Aquada more than 15 years ago, registering scores of patents covering all sorts of technical advances in the process. The car was powered by a 2.5 litre Rover V6 and could do over 100mph on the road as well as an amazing 30mph on the water. It's the car in which Sir Richard Branson set his 2004 record for crossing the English Channel in an amphibious vehicle. He did it in just over 1 hour 40 minutes, cutting more than 4 hours off the previous record. I was fascinated at the time and started following the company's progress.

Conscious that I hadn't heard anything about Gibbs for a while and, in January 2017, finding myself with some time on my hands in Nuneaton, I went and knocked on their door to see what they were up to. Well, we got talking, and after several hours we had reached the conclusion that a) they were going to lend me an Aquada for the summer and b) I was going to help them design an all-electric Aquada. Result!

I had a blast in the Aquada all summer. There are slips all over the place for launching leisure boats, which work perfectly for the Aquada too. As do any bank or beach that are stable enough. I also took part in the very select biennial 'Amphib' at Henley-on-Thames. It's an event organised by the European Amphibious Car Group and around 60 amphibians from all over Europe gathered for a week of 'swim-ins' and generally spending time with like-minded people. It's a close-knit group, held together by their shared love for, and ownership of, amphibious vehicles – including historic military vehicles and impressive (some quite terrifying) home-made efforts to a handful of Amphicars.

It was a great privilege to be invited to join in my Gibbs Aquada. We gave people rides on the river and there was an unintended comedy act by yours truly when I had to make several attempts at getting the Aquada up a temporary slipway out of the river. My two passengers and the people watching from the riverbank thought it was a great laugh and provided many helpful suggestions: 'Oi, try turning it off and on again!' I kept smiling but breathed a huge sigh of relief when I finally managed it, having drenched the closest spectators on the way. It's all part of the fun.

Having spent the summer messing about on the river, it was time to get down to the second part of my plan with Gibbs – developing the very first all-electric high-speed amphibian, which we named Equada as a working title.

This project was pure joy. I got to combine two motoring passions – amphibians and electric power – and had the time of my life. The Aquada uses a water jet for propulsion, rather than propellers like the 1960s Amphicar, and is a much faster, more complex vehicle. The wheels fold up into the wheel arches to make it more streamlined and allow the car to rise up on a planing hull beneath the vehicle, just like a speedboat, meaning that there is less of the car in contact with the water and less drag, allowing it to reach higher speeds.

The plan was to convert the Aquada into an Equada and take it to Coniston Water, where they hold a week of speed trials each year in November. Boats of all shapes and sizes – from long, sleek power boats to semi-inflatable RIBs – aim to set records in their own class over a set course. The faster power boats can reach over 100mph and we were hoping to set a record for fastest electric amphibious vehicle.

By the time we started work on converting the Aquada to electric drive, in October 2017, there was little chance of having it ready for the Coniston speed week in November, but we took the opportunity to do some testing with two of their other vehicles instead. I'm not the ideal build for a racing driver (more of a horse than a jockey, you might say), yet we managed to bag a couple of records for fastest amphibious vehicle in two separate engine size classes.

One was in the Gibbs Humdinga, which the company describes as an amphitruck. It's a three-tonne beast that can carry six passengers and plenty of baggage, so it had no trouble with me on my own. The Humdinga got to 39.6mph, setting a British record in the Amphibious Experimental Unlimited class. I was super-chuffed, until some young pretender called Jeremy Clarkson turned up later in the week with his *Grand Tour* entourage in what looked like a paper aeroplane, with the same engine power. Needless to say, he smashed my record with a time of 47.81mph.

Our second record was a bit more of a wind-and-water-in-your-face thrill-ride. As the name suggests, the Gibbs Quadski is a cross between a quad bike and a jet ski. On land, it is a go-anywhere, all-terrain, bugs-in-your-teeth joyride that can hit 45mph. On water it is just as quick. I got it to 48.51mph. That record still stands ... but the greatest honour was to be awarded a K7 Silver Star for having established two new records. This award dates back to Donald Campbell's famous Bluebird K7. The K7 club was formed by Donald for his friends who helped him with the record attempts. The club is now open for people who have taken a boat to over 100mph – who are awarded a gold star – or who have established a new record – gaining them a silver star.

* * *

The following year, in November 2018, the Equada was ready to attend the next Coniston Water event for some high-speed testing. There were of course a few teething problems – there always will be on a project like this.

Even our journey up to the Lake District didn't go as planned. I do a lot of miles every year, usually towing some project or other and, as Sod's Law would have it, my old Range Rover – a proper workhorse that has done over 240,000 miles – had sprung a leak. I spent ages the night before we left removing various bits from under the bonnet and spreading them out on the dining-room table to try to sort it out (Imogen loves it when I do that).

The problem turned out to be a leak in the thermostat housing – a really complex part that looks a bit like an octopus and is connected to just about everything. A new part was needed and, as there was nowhere to get it at silly o'clock in the morning, we ended up travelling to Coniston Water in the rather less luxurious VW work van.

Now, I love my T5, but I must concede it is a little grubby and full of tools and lots of 'vitally useful stuff'. Imogen was not impressed with our last-minute transport switch, especially as the van doesn't have heated seats and it was freezing and snowy. Still, it got us there safely and in relative comfort and it had to have been better than our other option, the Casual Lofa ...

The next morning, Coniston Water glittered in bright sunshine, but it was still bitterly cold. I'm not too bothered by a bit of bracing weather, but low temperatures are not ideal for getting the most out of batteries in an experimental vehicle.

The guys from Gibbs had made a few minor changes to the Equada since I had last seen the car, so we had a run through of those and then I drove it down to the shoreline where we were able to plug it in to warm the batteries before our allotted time slot. As I drove the car into the water to get in line for my speed attempt, the wheels folded up into the arches as they're supposed to, but the dashboard lights that were supposed to be lit to tell me everything was OK, weren't. A bit concerned, I backed up towards the slipway, and – of course – switched it off and on again. That did the trick; everything then appeared to be working as it should and I made my way out from the jetty to the marked course.

When we got there, the Equada behaved perfectly. I didn't want to push it too hard, but the 200kw motor had the car surging forward and up onto its plane in no time. I made it to a record 28.7mph, although, truth be told, because it's the very first in existence, I could have set a British record for a high-speed electric amphibian at just 2mph. The Equada is, literally, in a class of its own.

We had deliberately held back from pushing the car to its limits on the first run so I knew that I could get more out of the car as we had one more run booked for the following morning. Conditions were exactly the same, if even colder. The sun was bright and the lake was dead calm.

This time, however, when I drove the Equada into the water, systems seized up and it refused to shift into marine mode. After a couple of attempts, the wheels came up and the dashboard indicated that all systems were go, but at the same time, a lazy wisp of smoke curled slowly skywards from the back of the car. Clearly, we weren't going out on the lake that day.

Because of the cold, the hydraulic fluid was very viscous and it appeared that the hydraulic pump had been drawing too much current for our electrical system in its attempts to shift the fluid.

Having got the Equada back to the workshop and stripped it down, it turned out that the smoke was simply from one of the pre-charge resistors overheating, so we probably could have got away with doing the run, but when playing with high-voltage kit like this, you just can't take the chance.

Even so, we came away from the Lake District with a respectable new record and a bit more work to do. The Equada will be zipping across Coniston Water again at the next speed trials in November 2019. In the meantime, I can't wait to crack on with the work.

12.

ANCIENT AND MODERN

Before sunrise, on the first Sunday of every November, a tiny corner of London goes back in time to Edwardian Britain. Hyde Park Corner bustles with men and women dressed in tweed, fur and leather, in plus-fours, capes and caps. The foggy air is filled with steam and smoke and smells of soot and oil. The tranquillity of the park is shattered with the sounds of chuffing exhaust and whining gearboxes.

The start of the London to Brighton Veteran Car Run has such a wonderful, evocative atmosphere. It is the longest-running motoring event in the world and it celebrates the Emancipation Run when, back in 1896, at just past midnight on 14 November, the latest Locomotives on Highways Act came into force, raising the speed limit of 'Light Locomotives' from 4mph to 14mph.

The original version of this Act had been passed in 1865. It restricted any 'self-propelled road vehicle' to the equivalent of a fast walking pace and stipulated that someone must walk ahead at all times, carrying a red flag to alert other road users. In the 1860s, 'road vehicles' were likely to be traction engines. But by the 1890s, cars were becoming a more and more familiar sight on Britain's roads and the rules had become a bit of a joke.

So, to celebrate the change to the law, designer and motor industry pioneer Harry J. Lawson organised a 55-mile trip from London to Brighton on the first day British motorists were allowed to drive without having to follow a man on foot. Thirty-three cars turned up and the proceedings started with a symbolic tearing in two of a red flag by politician Lord Winchelsea.

I love the adventurous and pioneering spirit of the Victorian and Edwardian eras; it was an amazing and innovative time for early motoring, and what we now think of as our history was their cutting-edge technology. Nobody knew what these ground-breaking machines should look like or how they should work, so everybody just had a go and tried to solve the challenge their way. Anything was possible and everything was tried.

Back then, hundreds of engineering adventurers set off to explore unknown territory and some of their discoveries and inventions endure to this day. One of those pioneering engineers was Frederick William Lanchester. He was an inspired and prolific inventor who patented the first caliper-type automobile disc brake in 1902 and we have a lot to thank him for. For example, I was recently working on a 2018 Mercedes

Sprinter and noticed that even its modern engine had a pair of Lanchester Balance-Shafts, patented in 1904, spinning away deep inside, offsetting the vibrations from the crankshaft and pistons, providing smoother running in 2018. The de Dion tube (from the De Dion Bouton), invented 1894, and Panhard rod (from the Panhard Levassor), are also still in use in modern suspension systems today.

In 1900, 38 per cent of the cars sold in America were electric, 40 per cent were steam-powered and only 22 per cent had what they called 'exploding' engines – the internal combustion engines that most of us still drive today. In fact, to fuel these early internal combustion engine vehicles, pretty much anything that would explode was tried: coal, petrol, paraffin oil, acetylene gas ... One jalopy even used a mixture of moss and coal dust, and Rudolf Diesel's invention (the diesel engine) was being run on peanut oil.

So, in 1900, as well as electric vehicles being more prevalent, some of the first diesels were actually bio-diesels. It is amazing how it has taken 120 years for the tide to change back.

In 1897, a fleet of 8HP (6kW) electric taxis, nicknamed Hummingbirds because of the cute noise they made, operated in London. In 1898, long before Ferdinand Porsche had ever imagined the Beetle, he invented an electric front wheel hub-drive system, resulting in the world's first petrol/electric hybrid vehicle – the four-wheel drive Lohner-Porsche. His big coach was dubbed La Toujours Contente ('The Always Satisfied') a playful dig at another electric vehicle hitting the headlines at the time, La Jamais Contente ('The Never Satisfied') – Camille

Jenatzy's torpedo shaped racer in which he set the first ever automotive speed record in 1899.

Back then, battery cells were made from glass and were heavy and fragile; the Lohner's lead-acid battery pack weighed 1,800 kilograms and had a capacity of 21.6kWh. By comparison, through its use of modern Lithium-ion chemistry, a modern Nissan Leaf's battery pack weighs a sixth of that, around 300 kilos, and stores twice the amount of energy, around 40kWh.

Part of the problem for these ambitious late Victorian and Edwardian inventors who wanted to use electricity to power their vehicles was that, aside from inadequate battery technology, the cost of electricity at the time was around 68 times what we pay today. The generation of electricity was mostly a private affair, with individual companies feeding their local area. (The national grid wasn't implemented until 1937.) Mostly, electricity was created by fuel-burning generators, though the first hydro-electricity plant was built in 1878 for a country house in Northumberland. Even though solar cells did exist, they were in the embryonic stages of development and not available for anybody to stick on their roof at home. So, although at the dawn of the automotive era, electric (and steam) vehicles were by far the most attractive and preferred option, it's taken until recent years for technology to catch up to that vision. With easy, affordable access to electricity and viable battery technology we have been able to set out on new adventures in electric motoring.

* * *

Near the end of Series 11 we were due to reach our 100th car, and we wanted to mark the occasion by doing something a bit special. I was keen to find a vehicle that we would never normally get the chance to work on. A veteran car seemed ideal for a 100th celebration, and the London to Brighton Veteran Car Run would be a suitably challenging and spectacular test drive.

All we had to do was to find one that needed a little TLC, we thought. However, to be eligible to take part in the veteran run, we needed a car built prior to 1905, making it at least 110 years old. The tiny catch was the cost of such a car; you can sometimes find a bargain for less than £50,000 but the price could easily be upwards of £150,000. Not a chance on the *WD* budget. So, we needed to borrow one, and the search was on ...

We got in touch with the Royal Automobile Club in Pall Mall, Toby and Daniel Ward of the VCC (Veteran Car Club of Great Britain) Dating Committee, and Doug Hill of the National Motor Museum, Beaulieu. There were so many great cars and compelling stories, but they were either beautifully restored already or untouched original examples or else needed far too much work.

Eventually, we found the ideal project at the Haynes International Motor Museum. Anyone who has ever attacked an engine with a spanner knows about the Haynes Owner's Workshop Manuals. The company produces books that show you how to tackle all sorts of jobs on everything from a Morris Minor to the *Millennium Falcon*.

The museum, in Somerset, is an absolute treasure trove of classic, vintage and modern cars. The car that they were prepared to lend us was a 1903 Darracq. These were French cars, although the company was owned by a British consortium, and they had a solid, even glamorous reputation. Darracq specials held the World Land Speed Record in 1904 and 1905, clocking over 100mph. This 1903 single cylinder, 8HP engine wouldn't be going quite that fast, though.

Alexandre Darracq had started out manufacturing bicycles and, like so many companies at the time, went on to apply his engineering knowledge to automobile design. By 1904, his company was producing more than 10 per cent of the automobiles in France. In 1906, he founded a company that became Anonima Lombarda Fabbrica Automobili (A.L.F.A.), which in turn became Alfa Romeo in 1910.

Our little car was similar to the famous 'Genevieve', a 1904 12HP twin-cylinder Darracq that was driven in the London to Brighton rally in the 1953 movie of the same name, and is now owned by the Louwman Museum in Holland. However, our car hadn't been driven for nearly a decade, and so before it was fit to take part in the 2013 run it was going to need some work – which proved to be an exciting if not slightly daunting challenge.

Mike and I were filmed being introduced to our Darracq in the spot where it had stood on display for many years, tucked away in a corner of the museum opposite a couple of very fine classic Bentleys from the 1960s.

Outside on the Haynes track, we managed to eventually jump-start it while being towed. Coaxing the Darracq into life was a magical experience. It was a real thrill to be in a car that was more than a century old, even if it did feel like it was going to shake itself, and us, to pieces.

The first issue we looked at was with the wooden wheels. They were loose because they had dried out and the wood had shrunk. We tried plimming them – soaking them in water to make them 'plim', or swell, to tighten the joints. This wasn't hugely effective though, and we later discovered that really wasn't the right thing to do as the wood will swell when wet but then will simply dry and contract again. So I took them down to a royal wheelwright (the guy who makes and maintains the wooden wheels on the Queen's carriages) near Axminster in Devon to repair the wheels properly and then shrink-fit the metal tyre rim back into place. Watching such a skilled artisan apply this ancient knowledge was a real privilege – Greg had learned his craft from his father, Mike, and was already training the next generation.

Another of the jobs was a little surprising: we had to oil the clutch. It is a little counterintuitive, but it's because the clutch on the Darracq is a cone made of leather, and if the leather gets dry the clutch will snag. However, they are even more trouble if they get wet ...

Some years before, I was in another Darracq on the London to Brighton run, amidst torrential November rain. The clutch was getting harder and harder to disengage until finally we had to stay in gear and negotiate the traffic lights on the winding approach to Madeira Drive in Brighton fixed in gear. It turned out that there had been so much water on the road that the leather clutch had been absorbing moisture for most of the journey, expanding all the while, taking up any slack in the clutch mechanism meant for disengagement. Stopping was not an option and so we were forced to jump the lights, dodging pedestrians and motorists alike, in a rather hazardous hair-raising fashion.

I did once get caught on camera advising Wayne Carini of *Chasing Classic Cars* on Discovery that, as the Edwardians didn't have traffic lights (the first ever traffic lights were installed in Piccadilly Circus in 1926), Edwardian cars and their drivers should probably disregard them!

On our first test drive at Haynes, I stalled the Darracq many times, and then, once we got going, we lurched violently with each gear change. The dry leather clutch was snatching onto the metal flywheel, so by oiling the leather, the two moving parts slipped together nicely giving a lovely transition into each gear but still gripped without any slip once a return spring held them back together. A really simple solution, using materials that were plentiful back in the day.

As our Edwardian car was born of the cutting-edge technology of its time, I wanted to apply some modern-day cutting-edge technology to our refurbishment. The car sported a pair of lovely old Lucas 'King of the Road' brass coach lamps, but sometime in the car's murky past they had been a little butchered. Instead of using the flame from the original wick soaked in oil to light the way, someone had removed the lamp inserts and bolted in a pair of old incandescent motorbike lamps, powering them from a motorbike battery. It certainly made sense to upgrade the lighting, as driving on the road with nothing but a pair of oil lamps to light your way is downright terrifying. However, I figured we could do a lot better and still stay sympathetic to the originality of the car.

First of all, though, we restored the original lamps. We cleaned up the oil lamp inserts, removing all of the stale oil residue inside, threaded in new wicks and filled them up with fresh oil. The lamps now looked very smart and burned beautifully but the light they gave out was pretty hopeless, especially if you want to actually see where you are going on a dark foggy morning, or indeed, would like modern traffic to see you.

What I wanted to do was to make the lights work to modern standards, without it being possible to tell by looking at the original lamps. I started by measuring the oil lamp inserts and drew up a 3D model in my CAD.

In the same way that I was taught to use 2D AutoCAD at university, the Edwardian engineers who designed the coach lights would have trained as draughtsmen. Engineering drawings were done by hand on a drafting table and, once perfected in pencil, the design would be made permanent with ink. Producing a drawing in 2D CAD was very similar in process, except our digital image could be easily changed, mirrored, scaled and created in layers that could turned on and off for clarity.

My friend Marc Amblard is a brilliant engineer and he introduced me to a 3D CAD program called SolidWorks. It allows you to create simulated solid forms on the computer the same way you would make them out of lumps of metal in real life by cutting holes, extruding shapes, and cutting chamfers and fillets. The models are so accurate they have attributed physical properties like density and elasticity, and you can even create mechanisms with levers and pulleys and then try them out in a simulation. The software really helped to speed up my design process and let me try out all kinds of designs on computer rather than having to experiment in the workshop. The best thing is, over the last few decades, manufacturing itself has also become digitised.

In 1967, the first industrial laser cutter was invented and now even a high-pressure jet of water, containing tiny granules, can be used to cut metal, glass, wood, almost anything. But it was the ability to program a computer to control the cutting that really revolutionised manufacturing. Now, we can design and cut complex shapes and assemblies in a fraction of the time.

The first time I got to fully test this technology was when I was commissioned to build a tool-themed-racer for Silverline tools and Airwaves gum (coincidentally for a race at Haynes Museum). I built a car out of a shortened Beetle floorpan that utilised as many hand tools as possible in its construction.

I used a big toolbox for the seat, large ring spanners for the pedals, spanner rolls for the wheel arches, hand torches for the headlights and a compressor receiver for the fuel tank. It was a lot of fun trying to replace all of the controls and usual features with some kind of hand tool. When it came to the engine, though, I needed something a little special.

I took my inspiration from a very special engine designed by my friend Steve Prentice. It was an X12. As the name suggests, rather than, say, 8-cylinders arranged in a V, this was twelve cylinders arranged in an X. On paper, it was extremely powerful for its size and weight. I was creating a tool car, so instead of 12 cylinders I planned to use 12 chainsaws. Obviously.

I measured up the chainsaws and created a simple 3D model in my SolidWorks. I sourced some pro-kart sprockets and chains, a standard shaft and a selection of pillow bearings, digitising all of those by hand too. Finally, I measured the VW gearbox bell-housing, flywheel and clutch and once I had most of the parts in the CAD, I could start to assemble them on the computer.

I started by working on the first bank of four, to make sure the position of each chainsaw was going to work, and then designed a replacement sprocket to connect each chainsaw to the central shaft. Once I had worked out how one throttle cable could control all twelve chainsaw triggers I literally copied and pasted the first bank of four and turned my X4 into an X12.

After that, it was easy enough to join the dots – quite literally – and create flat metal forms that would join everything together. It took a couple of days to finish the design and then I simply emailed the drawing files to a local laser profiler. A day later I picked up a box of really cute metal parts. They were about a quarter of the size I was hoping for. I had forgotten to include any dimensions for the scale. It was all a little bit like the Stonehenge in *This Is Spinal Tap*.

The next day, I collected a bigger box of parts, all the correct size, and a day or so after that I had welded up my engine frame and was bolting on sprockets, chains and chainsaws. It was amazing that in less than a week I had gone from a blank screen, to a 3D model, to a fully assembled engine.

The plumbing and cables took a few days more, as they always do, but when it finally fired up it sounded fantastic, like a squadron of Messerschmitt. We came second in the race but frankly it was brilliant just to be driving around a track powered only by hand-tools. What if I tried bigger chainsaws and more of them? X36 anyone?!

Anyway, back to the Darracq headlights. Having made the 3D model of the headlights' original oil lamp insert, I then measured a battery holder and bulb holder and an LED headlight bulb. I modelled these up in SolidWorks too, and modified my model of the oil lamp to accommodate them. In a few hours, I had a 3D representation of my LED alternative for our light source. All I had to do now was to make it.

In the mid-eighties, the very first 'additive manufacturing' process was developed and 3D printing was born. Now, shapes could be created directly by laying down material one pixel at a time, like hundreds of layers of 2D printouts cut out and stuck on top of each other.

At first, the process was very expensive and the materials were too brittle to be of any practical use, but over the last few decades, simplified machines have become very affordable, even available on the high street. In addition, the range of materials available is now huge and they are strong enough to be used – with a little finishing – straight off the machine. In short, it is now possible and commonplace to print with all kinds of plastics and rubbers, but also metals, even steel, stainless steel and exotic metals like Inconel, which is often used in motorsport and for aerospace.

When the engineers who were working on the Darracq at the turn of the last century wanted to cast a new metal part, they would have first needed to make a wooden pattern or model of what they wanted. The pattern would then be sandwiched between two halves of a mould made of compacted sand and, once the pattern had been carefully removed, the resulting void could be filled with molten metal.

It is now possible to print the sand directly into the mould, creating complex voids that would never have been possible before. Additive manufacturing is revolutionising the world and we have only just started exploring the possibilities.

To make our new lamp inserts, we were able to borrow a high-end 3D printer that could mix two materials in any proportion – in our case a plastic and a rubber. We chose a setting that would be rigid but flexible enough to damp some of the vibrations from the car and loaded my drawing file into the machine. Then all we had to do was set up a time-lapse camera, press print and leave the machine running overnight. Eight hours later, we had two identical LED coach lamp inserts.

They worked brilliantly; the LED light bulbs gave out a strong, white light, and they consumed so little energy that one set of batteries would last nearly a week. We even started a trend.

A friend of mine, Henry Lawson, was campaigning a pair of veteran MMCs on the run. His son Rowan, inspired by my conversion, digitised and printed his own LED inserts for the various lights on their MMCs. I recently found out that the very same Harry J Lawson who organised the original Emancipation Run, not only owned the Motor Manufacturing Co. of Coventry that built Henry's MMCs, but was also Henry's great-great-uncle!

3D printing is revolutionising the restoration of vintage and veteran cars, particularly when there are only one or two examples remaining. Normally that rarity means there are no spare parts available anywhere so they have to be made from scratch, from plans if you are lucky, or just by working with whatever remains.

American comedian Jay Leno, to whom I was introduced by another car collector friend, Bruce Meyer, has a wonderful collection, including a glorious brace of Duesenbergs. Parts certainly aren't readily available for those hand-made, custom built beauties, but he and his team have a 3D printer in their workshop. So, when the water pump on one of the Duesies failed, they were able to scan it into a computer, re-engineer the virtual pump to remove the signs of damage and corrosion and then print out the part in plastic to test the fit on the car before eventually committing to having the replacement part made in metal.

* * *

Getting our car in good running order was one thing, but driving the Darracq was a complex affair that required huge concentration – although it was also huge fun. Because nothing is automatic, simply starting it is quite a performance. The timing of the spark from the single plug in the cylinder needs to be carefully adjusted. On later cars with distributors, this is done automatically, but, on our car, the timing had to be retarded for starting, slightly retarded going uphill and advanced for most normal driving.

There was a lot to think about – as well as the timing of the spark, there's the fuel/air mixture, the throttle position and the gears, as well as steering the car. It was like spinning plates while juggling crystal decanters. I had spent an afternoon training on the car with Nigel Parrott of Veteran Engineering. He and his team maintain many of the Veteran Run cars and he gave me some great tips but now it was all down to me.

On the day, the weather was perfect for our London to Brighton run; through the mist of a cold and crisp November morning we trundled through the staging lane in Hyde Park, cosily cossceted in tweed, our LED lamps ablaze. The crew fitted the cameras, Paul and Phil did some last-minute checks, we did a small piece to camera and we were off. Only to break down just past Buckingham Palace.

I pulled out the sparkplug, and cleaned off a load of soot. We were soon on our way again, relieved things were going to plan. Progress was good, we had captured some great shots and despite getting very, very close we hadn't actually hit the tracking car. We stopped off at Crawley for a cup of tea and a `tech check' of the filming equipment and then cracked on, towards the dreaded Clayton Hill.

A little further on, the oil lubrication pipe fractured because of the vibration of the engine. We made it into a road-side car park and, after a bit of bush mechanics, and a sandwich, we were back on our way to perhaps the biggest challenge of the run.

Clayton Hill can make or break a run: over the short distance of less than a kilometre, the hill climbs by 60 metres with an average gradient of 6.8 per cent (1:14.7) – the steepest gradient being 13.6 per cent (1:7.3). Pretty much everyone we had spoken to in preparation for this run had warned us about this part of the rally.

To give us a fighting chance, I got up as much speed as possible on the approach – whatever happened, we mustn't stall the car. The organisers close the road to traffic coming down the hill and local volunteers wait patiently with 4x4s and tow ropes. In the past, as a passenger, I had seen veteran cars go up the hill three abreast in fear of a car in front of them stalling or stopping and causing them to lose valuable momentum.

As we climbed the incline, our plucky Darracq started to slow. I changed down to first, but we were still losing too much speed. I ordered Mike to get out of the car and run alongside. It helped, but the car was still struggling up the steepest part.

There was only one thing for it: I jumped out too, and ran alongside – the throttle on the steering column was already set and a little light steering adjustment kept us out of the verge. That definitely helped, but how far were we going to have to run alongside? As we neared the top, the car finally conked out – we were so close! Thankfully, a little gentle persuasion got the car going again and we trundled on our way to Brighton. As we dropped into the city and past George IV's Royal Pavilion, we knew we had done well.

We spluttered over the last roundabout and onto Madeira Drive; we could push it over the line from here if we had to. The emotion started to well up as we made it over the line – we had completed our epic adventure. It had taken us a day to travel the distance a modern car could manage in less than an hour, and what a wonderful experience it had been.

* * *

Many classic cars were originally designed to last for around 100,000 miles. Even assuming a meagre 5,000 miles per year, that only gives them a 20-year life span, so any classics still around now have certainly earned their keep and the number of classic cars that are still on our roads is a testament to people's love for them.

Of course, as all cars get older, they get harder to maintain and as their performance and reliability increasingly falls short of the standards we are used to in modern motoring. Even if there are people interested in maintaining them, they're still in danger of becoming irrelevant and eventually obsolete, even the next environmental pariah.

Thankfully, the resurgence of electric propulsion might have an answer – electrification of classic cars. We explored that option in one episode of *Wheeler Dealers*. For some six years I had been pushing to do an electric conversion on the show and when we moved to California, the stars finally aligned.

We were putting together our list of car projects for the series and Henric Nieminen, our 'Finnish and Fantastic' studio director, had a lead. Early Maserati Biturbos, with carburettors rather than fuel injection, were notorious for their lack of reliability. In 1993, Henric's father, Martti (or Mara for short), had removed the defunct 2.5-litre twin-turbo V6 engine from a 1985 Maserati Biturbo and replaced it with a cleverly modified electric drive system from a forklift truck. The car had been sitting in a garage in Brisbane, California, for 15 years and the idea was to upgrade its very dated technology with a modern electric drivetrain and battery pack.

We got in touch with Michael Breem from EV West, a visionary company based in San Marcos, who have been converting classics to electric drive for years. Michael's own daily driver is an electric VW split-screen pick-up. He has installed a battery wall and solar panels on his workshop's roof, so now he gets to drive around for free. Admittedly he is guaranteed rather more sunshine in Cali than we are in Blighty but it is still just as workable and that project is also on my list.

Under the guidance of Michael and his team, we set about doing a modern electric conversion. First, we removed the old motor and all of its ancillaries, and then we digitised the empty engine bay and gearbox to create a virtual model in SolidWorks.

Because our Maserati had been a 'Biturbo' I thought it would be fun to go with a 'bi-motor' set up, so Michael suggested we use a 'siamesed' design with two AC brushless induction motors fixed to a common shaft. Using the virtual model in the computer, we created an adaptor plate that mated the electric motor onto the Maserati gearbox. The lithium batteries were salvaged from a modern electric Smart car.

When converting a classic to electric drive it seems obvious to me that the starting point should be trying to better the original performance spec of the vehicle, but, crucially, the work should be fully reversible. Ideally, the original engine should be restored and either put in a crate for safekeeping – or perhaps made into a coffee table for the workshop! Either way, the car's history should be preserved.

A great feature of an electric conversion is of course that you can 'fill up the tank' at home, so your classic will always be ready to go when the whim takes you and it will start on the button, each time. You also get to choose where that power comes from – from the grid, your own solar panels or perhaps your own windmill. And you get to change your mind. If, a couple of years into owning the car some new way of generating electricity is invented, you still fill up the same way, just from a different source. For our test drive we visited the amazing Crescent Dunes Solar Thermal power station near Tonopah, Nevada. It's rated at 110 megawatts and can store 1.1 gigawatt-hours of energy. That could re-fuel a lot of converted classics.

Once we had finished with our Maserati, the new lithium battery pack gave us a projected range of 120 miles and the new performance specs were certainly an improvement on when it drove out of the showroom in 1985. Our electric set up gave us 142HP and 240lbf-ft of torque. That was slightly less than the Biturbo's horsepower rating when it left the factory, but 50lbf-ft more torque than it ever had.

The torque is important because an electric motor delivers

all of that torque from the instant you press the accelerator, whereas an ICE will produce peak torque only when it is operating within its optimum rev range. In practice, that means that even if the electric motor has the same headline torque figure as the original internal combustion engine, the electric conversion will have better acceleration.

The Biturbo came out of the factory in 1985 able to reach 60mph from a standing start in 8.4 seconds. More than 30 years later, on electric power, we did it in 8.1 seconds.

* * *

I guess I have a soft spot for all kinds of trailblazing engineers, whatever their era, whatever their frontier, passionately messing about in sheds with their bicycles, cars, amphibians, aero-engines, electrification, hydrogen fuel cells, rockets, photon-drives ... Thanks to today's visionary engineering adventurers and scientific pioneers we get to enjoy a constant stream of technological advances and these will have a continued impact on how we live and travel.

Take, for example, my dad's Ariel 3 satellite. When he was designing it in the late 1960s, the most advanced solar panels might manage to convert into electricity only 10 per cent of the energy from the sunlight that hits the panel. That was an enormous improvement from the 1 per cent efficiency of the solar cells that had been introduced way back in 1883. Now, we can buy a relatively affordable panel that can muster 21 per cent, and if you are very rich or building a spacecraft with someone else's money, you could get your hands-on today's cutting-edge technology cells with an efficiency of 37.75 per cent. Progress like this brings with it a raft of new possibilities – our planet is solar powered, after all!

Our world is shaped by pioneers of all kinds, those who dare to run headfirst into the unknown with unbridled joy, like toddlers and dogs when they see an open space. These people need to be cherished, encouraged and celebrated. Because of these curious, unreasonable people, our world is a better place and our lives more interesting and convenient.

Here's to embracing all kinds of pioneering spirit, ancient and modern ...

13.

BACK IN THE DRIVING SEAT

There's nothing like taking a spirited, open-air drive through the English countryside. It might not be the most dramatic scenery in the world, but I enjoy bowling along past grassy banks, ditches and hedgerows, enjoying the views of fields and woodland, rivers and streams.

One particular winter morning, while partaking in this innocent pleasure, I was enjoying myself immensely. The hedgerows were bare and the scenery was more a uniform grey than a lush green, but the weather was fair and the air was crisp. Well, actually, it was bone-chillingly cold.

A fine day for a brisk walk to a country pub to warm yourself in front of a roaring log fire, maybe, or to cruise the B roads in the pampered, heated luxury of a Jag or a Bentley. Unfortunately, the car we were in could not be described as luxurious and instead of heating we had icy wind blasting in our faces – there was no windscreen. It was, however, a very special car and a delight to drive.

I was in a 1907 Itala, clattering around the countryside filming part of the *Wheeler Dealers* 1916 Cadillac special with Mike in the passenger seat. We couldn't drive our own Cadillac because it was in bits in boxes and, truth be told, in a sorry state.

But that, of course, was the whole point of the show and, rather than regarding our 1916 Caddy as a complete basket case, I preferred to view it as a work already in progress. Much of the dismantling that needed to be done had, after all, already been done; some effort had been put in to bringing the bodywork together and the engine was with a specialist undergoing a rebuild. I like to think that we had hit the ground running on what was going to be an epic project.

Mike felt pretty much the same way, although he wasn't enjoying our icy country ramble quite as much as I was. Freezing your nuts off in a vintage car in winter is not everybody's cup of tea and Mike had been lured away from sunny California to make this film. On camera, he was playing the disgruntled doom-monger to my ebullient enthusiast. He was wearing an extremely grumpy face and I was grinning like a kid in a toyshop. Sure, it could have been warmer – but you can't have everything.

On the other hand, it would have been nice to have had brakes. The Itala is a lovely car with a massive eight-litre engine that produces 40 horsepower. That's far more than the 8 horsepower produced by our 1903 London-to-Brighton Darracq and it gives the Itala a top speed of around 70mph.

To put that in perspective, a modern Fiat 500, for example, has a 0.9-litre engine that produces 85 horsepower and a top speed in excess of 100mph – though it did take a century to get cars that efficient. So the Itala was a cool car for 1907, but its brakes were pretty crap by today's standards, making that 70mph top speed for none but the brave.

In addition, its cart-spring suspension and the artillery wheels meant hitting a pothole would have made the car lurch and buck so much that Mike would probably have been ejected from the passenger seat. Which meant that we were using more than our fair share of the road as I was steering around all the potholes on the country lane we were travelling along.

When filming a road test, the closer you bring the car to camera, the more dynamic the shot looks so you spend most of your time 'worrying' the back end of the camera car, rushing up behind it so that you fill the shot, then dropping back a bit or overtaking. You have to do that time after time in order to get the best possible shots. We had a few narrow squeaks, coming face to face with oncoming cars around a bend. At one point, encountering a learner driver coming round a corner as we were doing a rush-up in the middle of the road, I nearly parked the Itala in the boot of the camera car, making Mike lay an egg, much to the amusement of the crew.

Mike and I had been invited to compete in the Endurance Rally Association's sixth Peking to Paris (P2P) Rally and we jumped at the chance. There were two perfectly sane (in my mind) reasons for doing this. The first was that rally-prepared cars sell for more than standard restorations, so we could have a *Wheeler Dealers* happy ending after the event. The second was that the event would be an amazing journey through China and Mongolia into Siberia and, from Russia, on to Belarus, Poland, Slovakia, Hungary, Slovenia, Italy, Switzerland and France.

Preparing the car seemed to be a perfect series finale for *WD*, plus we could film a 'special' or even a mini-series travelogue about our perils and pitfalls during the journey and the people and places we encountered along the way. We would be able to get a lot of TV material from a 36-day, 9,000-mile journey.

Without a doubt this was an enormous undertaking, but Mike and I are both huge car enthusiasts, and car enthusiasts had been taking part in the Peking to Paris for years. Plus, in 2007, which was the 100th anniversary of the original event, 126 cars had set out and 106 finished, so the odds of us making it all the way through to Paris were actually pretty good. It was a tremendous idea, would make wonderful television, and we would have the experience of a lifetime.

The TV execs at Velocity had agreed to the P2P project months before, but were clearly not as keen on the adventure as us, and were now employing a strategy of delaying decisions and budget sign-offs in the hopes that we would just get fed up and give up on the whole idea. However, as they were contractually obliged to let us do this project, Mike and I were not going to let them play that game.

However, any delays would give us less time on the project, so we decided to dip into our own pockets to buy a car rather than waiting for them to get their act together. We looked at a number of vehicles before some friends introduced us to a guy called Roger, who had an unfinished rally car project. It was a 1916 Cadillac Type 53 with a 5.15-litre V8, producing 77 horsepower – in fact, the first ever production V8, originally introduced on the 1914 Type 51 model.

A 'pioneer' car like this seemed perfect, absolutely in line with the spirit of the original event. Sure, right now, it was more a pile of bits than an actual car, but that gave us room to make it our own; we needed to upgrade the suspension and strengthen and protect bits that were likely to take a battering on rough roads anyway. It was all part of the challenge.

Having paid £37,000 for some beautiful Cadillac parts and a ton of rusty bits, we used the Itala joyride to introduce the idea of the Peking to Paris Rally, an epic driving adventure, testing the mettle of man and machine.

A 1907 Itala identical to the one we were driving that winter's day had won the very first Peking to Paris challenge. The race originally came about to counteract a tide of negativity against the motorcar sweeping through Paris at the time. People were complaining about these noisy, smelly, scary machines that kept breaking down, spilling oil in the street and frightening the horses – the general consensus being that they would never catch on.

So, in order to prove that the motor car was, in fact, a useful machine with a great future, the Parisian newspaper *Le Matin* threw out a challenge in January 1907:

> What needs to be proved today is that as long as a man has a car, he can do anything and go anywhere. Is there anyone who will undertake to travel this summer from Peking to Paris by automobile?

Five vehicles turned up at the French Embassy in Peking in June of that year to take the challenge. One of those was the 1907 Itala driven by Prince Scipione Borghese, an Italian aristocrat, sportsman, politician, industrialist, adventurer and, by all accounts, a bit of a cad.

Prince Borghese pulled a number of slippery, Dick Dastardly-style tricks during the race, including taking a shorter, direct route by having his Itala dismantled and carried over a mountain pass by a team of Sherpas, and booking out all of the hotel rooms at the various stopovers, so the other competitors had to find somewhere else or rough it under canvas. He also had his chauffeur with him to do the driving whenever he started to find it a bit too tedious.

Unsurprisingly, Prince Borghese came first in the race. His reward was the glory of victory and a bottle of champagne.

This race, and others, proved that the automobile was more than just a passing fad. In fact, by 1907, manufacturers were using races of all descriptions to prove the reliability of their cars, but the Peking to Paris was an immense challenge that would not be run again for 90 years.

Then, in the 1990s an exuberant chap and motoring enthusiast called Philip Young happened upon a 1964 book, *The Mad Motorists* by Allen Andrews, which was all about the 1907 race. Philip was instrumental in re-establishing the race and founded the Endurance Rally Association. The ERA's Peking to Paris was staged in 1997, again on its centenary in 2007 and every three years since. Sadly, Philip died following a motorcycle accident in Burma in 2015 but the ERA continues to organise epic rally adventures all over the world.

So, having read *The Mad Motorists*, I was thrilled to be thundering along in an Itala that had been restored to the specifications of Prince Borghese's car. We were all set to take part in the 2016 event but hurtling along through Home Counties' country lanes in the icy cold weather, it started to dawn on us why they called those motorists mad to take on such an epic journey in boneshaking jalopies.

We still had a TV series to make, of course, and we ploughed ahead with that while the Cadillac we'd bought was packed up and shipped across the Atlantic to the *Wheeler Dealers* workshop in Huntington Beach. We left the engine behind in the care of the specialists who had started the rebuild, so they could complete the work. However, when I went back several weeks later to organise filming the completion of the rebuild, I found that nothing had been done. I packed it up and shipped it to California, but when we finally did get our hands on the engine, it was clear we couldn't get the rebuild done in time. We decided we needed a donor car and, amazingly, we found the lovely 1918 Type 57 Cadillac in brilliant condition.

Our 1916 rallying Cadillac should have taken two years to rebuild, but we eventually squeezed the whole restoration into about 15 weeks. That was something of a miracle. I had help, of course, and lots of advice, not least from Jay Leno.

Jay has a Cadillac or two in his immense collection, including an unrestored and original 1918 Cadillac Type 57 Victoria that is similar in most respects to our 1916. He was kind enough to let us look over his car and gain all sorts of tips, and insights. Among our pile of rusty parts, we had found drums and brake shoes and a few other brake-looking parts.

While looking at Jay's car we realised that the Caddy used brake shoes on the inside of the drum for the emergency brake and another pair of shoes on the outside of the drum that were the actual service brake. We had only installed half of the braking system!

Jay also warned me about the pair of water pumps bolted to our engine block. They look symmetrical, and can be fitted on either side, but are actually handed so could easily be fitted so they counteract each other rather than work together to pump coolant around the engine. Jay and his mechanics had found this out the hard way but had noticed before any damage had been done. This proved a real time-saving tip.

Preparing the car to be rally-ready was one thing, but some purists might have objected to the way that I revamped the seats to have their own suspension and heating elements. This was a sensible addition to protect both my back and Mike's. There was no need for either of us to be crippled by the gruelling journey or to put up with the excruciating cramp I'd experienced in the Darracq on the way to Brighton. If we were going to drive for 9,000 miles, we needed decent seats with plenty of leg room.

My vision for the build was to create an Edwardian rally car as if I were an Edwardian gent doing the modifications using anything that would have been around at that time. The seat-heating elements were definitely modern but were easily hidden under period-correct upholstery. We also hid the compulsory GPS tracker in the dashboard and had the bezel of our Gauge Pilot navigation computer custom-made in brass to match the other dials.

I designed the dashboard in SolidWorks and had it laser cut, but ensured it matched that of the 1918 Cadillac so it felt correct. I also hinged the dash, providing a kind of glove box where we could keep important papers and also hide the camera and sound equipment from view.

The seats were mounted to a hinged frame so we would have easy access to the batteries sitting beneath. To keep things light and avoid redundant equipment, I gave many parts more than one function: the hinges for the seating were spare cart spring hangers for the suspension; and to dampen where the seat frame sat on the seat's suspension mechanism, I sleeved the steel tube with rubber hose the correct diameter for the cooling system. That way, should any of the hoses be damaged we would have spares to hand.

The springs I used for the seat suspension came from a 1907 tractor and the additional air horns I added to the front of the car came from a 1911 locomotive. They looked great and were immensely loud so they added character while being effective protection from wayward traffic.

The Cadillac's engine has a pump to pressurise the fuel tank rather than using a mechanical or electric fuel pump. As standard, it has a compressor running off the back of the engine so you can also pump up the tyres. This is a luxury that most cars don't have now but it made sense to make the most of that feature. The vintage horn came with an air reservoir so I hid that under the Itala-style fuel tank that came with our car, so we could use our augmented compressed air supply for cleaning off the fine Mongolian desert sand from sensitive equipment.

I was fascinated to discover that, out of the factory, our Cadillac came with a ceramic-coated exhaust system. It is something we think of now as exclusively for the motorsport world, but Cadillac knew what they were doing and lived up to their claim of being the 'Standard of the World'.

To increase engine performance, to add further character to our car and to add a way of heating the cab on frosty mountain mornings, I commissioned a tubular manifold that routed the exhaust out of the bonnet and down each side of the body like racing cars of the period. My cunning plan was to make a canvas tonneau cover for the cab, so the heat from the silencers would waft into our leg room.

I managed to source some aluminium SIGG water bottles that were centenary replicas of their original drinks-containers from 1908. The plan was to create Edwardian 'cup-holders' in the wooden base of the seats. The standard horn that came with the car was electric but single tone. I managed to find a claxon from a vintage Ford that fitted inside the casing perfectly. Hot-rodding the standard horn seemed very appropriate.

The roads and tracks of Mongolia can be very rough on a car, particularly after the rain, when huge ruts and pot-holes can form. In the past, many cars have suffered from stress fractures and breakages in the metal chassis and suspension parts. I replaced the fragile standard truck battery with four robust 'off-road' batteries. My idea was to run them in parallel for normal running and then should we, or another team, suffer a breakage we could use them in series to produce a high enough voltage for some emergency 'bush-repair' arc welding.

The Cadillac is a fascinating machine with so many innovative features that paved the way for modern motoring. The layout of the 1914's controls defined the position of what we think of as normal in a car today. It brought the gearstick and handbrake inside the car and positioned the clutch, brake and throttle pedals in the 'correct' order on the floor. It used a locking ignition key and was the first production car to be fitted with an electric starter. Having seen what Cadillac were doing, Austin and many other manufacturers copied many of the ideas and so they became ubiquitous.

It was Henry Leland, who bought Cadillac from a struggling Henry Ford, who insisted that all Cadillacs should have electric starters. A friend of his had once gone to the aid of a motorist who had stalled a Cadillac. Unfortunately, the motorist had not retarded the ignition properly, so when he cranked the handle it kicked back, hitting him in the face and breaking his jaw. He later died from complications.

Leland was heartbroken and vowed that it would never happen again with one of his Cadillacs. Which, for us, meant that messing about with the starting crank was not going to be one of our problems out in the wilds of Mongolia, although it made sense to carry one, just in case.

There were many things that needed some final adjustment, but the car was finished enough for our first 'shakedown' run, which was to be filmed for the end of the episode. We had the roads, tracks and beautiful rugged countryside of Tejon Ranch in Lebec all to ourselves. We took to the road in our rally-ready Cadillac for the very first time ... and, within 20 metres, ground to a halt.

In the previous series, we had modified the brakes of a 1963 Volvo PV544 with a special friction material, and as we only had brakes on the rear axle of our Cadillac, I figured the same modification would be invaluable to beef up our braking capabilities. However, the new material was so good at its job that it was binding almost immediately, so the shoes needed adjustment. I grabbed our on-board tool kit and freed off the brakes by the side of the road.

We filmed most of the test drive in torrential rain, which was a better test of the car than coasting along in perfect, sunny, Californian conditions. This was becoming the ideal practice run, fine-tuning our car in typical rally conditions. I had to iron out a few more problems throughout the day, but the car survived some rough driving over unmade roads quite capably, and made it safely up and down some pretty reasonable inclines. I was gaining confidence in our prospects for the challenge ahead.

And then the channel pulled the plug on the whole thing.

Towards the end of the day on our shakedown run, we got a message that we were required on a conference call, so Mike, his wife Michelle, Imogen and I piled into a car for the call. The production management informed us that they had decided that it was too much risk to allow their two valuable assets, Mike and I, to partake in such a dangerous activity and felt it 'essential that we act with an abundance of caution'. They had concluded they had 'no choice but to withdraw our participation from the 2016 Peking to Paris event'.

People have often commented about the rather strange ending to the Cadillac episode, where Mike is heard on voiceover saying that 'despite our best efforts' the car wasn't ready in time. Mike also says in the voiceover that he was sure he would be able to find a buyer for both of the Cadillacs. I had no idea that they were ending the show that way, but he did indeed find a buyer without too much trouble – me.

I bought Mike's share of the cars to own both Cadillacs outright. There is plenty of tinkering left to do on my rally car but I am already looking forward to, one day, campaigning the 1916 Cadillac in the Peking to Paris.

We finished filming the remainder of the series and then, months later, after 13 years of *Wheeler Dealers*, when Velocity created the unexpected opportunity to do so, it was time for us to part company. It was a huge change to my life and a lot to cope with. Like so many times before, I don't know how I would have got through it all without Imogen.

After the initial shell-shock had subsided, I found there were loads of new things that I wanted to explore and get stuck into. I've never had to look far for a car project to get on with. There has generally been more than one sitting just a few feet away on the driveway.

So, now that I had some time for forward planning, I thought of the long, long, ever-growing list of cars I have always wanted to build; I could finish the restoration of my Outspan Orange, develop a real James Bond Lotus submarine or get on with re-working the Ed Straker car from the '70s *UFO* TV series ...

There are also plenty of Cummfy car ideas that never made it to fruition: a driving beach, a rotating glitter ball, a dining table and chairs ... I also have the giant shopping trolley to revisit. The original electric motors were really not powerful enough for road use and the lead-acid batteries were too heavy, but I have now sourced a pair of very sporty motors and some much lighter lithium-ion cells, so maybe that should climb up my to-do list?

The Casual Lofa turned 20 years old in 2018, and I had always said I would rebuild it one day. Maybe the twenty-first anniversary would be a good moment to re-imagine what a twenty-first-century sofa-mobile should entail? Obviously, I have already given this a lot of thought. It needs to be electric, maybe four-wheel-drive, perhaps amphibious; it needs to have drinks fridges in the arm rests, storage spaces for waterproofs, a longer wheel-base, a waterproof stereo and absolutely needs to be fast enough to get back my Guinness World Record.

My trouble is, as ever, as soon as I get busy on any unfinished or even un-started projects, I find that solving the problems in front of me only gives me more ideas. Even writing this book has inspired many future projects – writing about electrifying classics gave me a new idea for a cool coffee table, writing about my land yacht gave me an idea for a designer chair.

Thinking about our RV led me to ponder an all-electric RV, which led on to a Controller Area Network (CAN bus) control for a house. Thinking about tsunami-proof housing spawned a new design for a studio workshop, and reading about Buckminster-Fuller's floating geodesic city inspired a design for a solar-powered airship. So, in a good way, one problem leads to another.

* * *

I was having a chat about attempting another world record with my friend Craig, the editor-in-chief at *Guinness World Records*. We went through my list of suggestions and Craig and his team wanted my thoughts about ... building a very fast ice cream van.

After a bit of investigation, it turned out that the current record is 200mph. The record was set by a Ford Transit-based ice cream van powered by a Jaguar XJ220, a legacy project from when Ford owned Jaguar, and is owned by former Formula One champion, Jody Scheckter.

Clearly, our budget wasn't going to stretch to a supercar powerplant like the XJ220's twin-turbo V6, so if we wanted to do a fast ice cream van record, we had to come up with a different approach. My first notion was to build an electric ice cream van. Everybody quickly agreed.

However, having said yes, I realised that once I had swapped the diesel engine for electric, I would no longer have the power-take-off for the ice cream machine. So, my second problem was: how to make the ice cream electrically?

My first stop was Whitby Morrison in Crewe, the world's leading manufacturer of ice cream vans. I had worked with them years ago when we created a couple of ice cream vans for Innocent Drinks and they had also converted Jody's Jag-powered Ford Transit platform into an ice cream van.

Back in the day, early ice cream vans had one engine to get them to their pitch or around their round, and a second that powered the ice cream machine itself. In the 1960s, Bryan Whitby invented a power-take-off system that allowed soft serve ice cream machines to run on the same engine as the van. This system revolutionised the industry and is still in use in pretty much all ice cream vans today. Whitby Morrison, now run by Bryan's son Stuart and his sons, confirmed that an electric ice cream van was sorely needed and they were up for being part of the project.

Whitby Morrison got busy converting a Sprinter, in double-quick time, into their latest style of ice cream van, the Amalfi. I went for a salt-flat racer inspired paint job using a retro orange, with cream scallops bleeding back from the front. I even persuaded their pin-striper to hand-paint the usual ice-creams on the front but have them splattering up the bonnet because of the speed.

As soon as it was ready, I dragged the van back to the workshop and got to work pulling out the brand-new Mercedes engine. It did feel a little odd to remove an engine that had only covered 14 miles, but the joy was, it being so new, everything was really clean and undid easily. I am sure it will be useful to someone. The fuel tank, exhaust system, ad-blue, turbo intercooler all came out – in fact, pretty much everything except the gearbox and radiator.

I then created a model of the gearbox bellhousing in SolidWorks, and of the electric motor, so I could start working out how to design some custom-made adapter plates to connect the two together.

I had elected to keep the gearbox, which is not strictly necessary with an electric motor, as it has such a broad band of torque. However, in this instance, being able to change up through the gears gave me an opportunity for better acceleration, but it also meant I needed a clutch.

The flywheel on the 2018 Sprinter is of a dual-mass design, which means it is effectively two lumps of metal joined by springs and dampers. It helps the engine run smoother and aids smoother gear changes, but I needed something I could machine more simply. Thanks to Mercedes' great legacy design, I managed to source a solid flywheel from an earlier model Sprinter that still used the same bolt patterns for the clutch and the crankshaft.

Guinness World Records had defined some rules for setting the record. To prove it was a proper ice cream van, we needed the chimes, the price stickers on the window and to be able to make soft ice cream before and after the attempt. To make the first record meaningful we also had to achieve a speed of at least 70mph.

This might not sound very fast, but as ice cream vans have all the aerodynamic properties of, well, a brick of ice cream, they are inherently rubbish at going fast efficiently. We did, however, manage to get to 79mph on our first proper test. The tricky part will be to do that on a measured course in front of official *GWR* adjudicators. Wish me luck!

Converting the ice cream van to electric drive was pretty easy. As I had anticipated, the difficult, second problem was how to make the ice cream without the power from the diesel engine. After all, it was a fairly critical part of setting the record that our ice cream van should actually be able to make ice cream. As it happened, though, without even intending to, I had somehow set out to fix a problem we have all suffered from since the inception of the driving ice cream van.

We all remember the sweet, sickly smell of ice cream mixed with diesel fumes when we were queueing up outside one of those vans, but I guess none of us really appreciated, or cared, that we were having our Mr Whippy served up with hydrocarbon sprinkles. In the modern world, however, diesel is seen as the fuel of the devil and we are all very much aware of how harmful the fumes can be, especially to small children.

Local authorities are beginning to clamp down, as churning out fumes beside a children's playground or on the beach is no longer acceptable and ice cream sellers are in danger of having their vans banned from their most profitable pitches. Whitby Morrison had been working on a conversion that allowed the ice cream machine to be run from a mains supply, which was fine for vans with regular pitches with access to power. However, to be truly mobile, a mobile solution was needed.

Ice cream men and women put in long hours in their vans during the summer months and the van's diesel engine has to run all day in order to be able to provide the cones of ice cream everyone loves.

The soft ice cream begins as a liquid mix, which is chilled in a hopper and then frozen and whipped on demand, and that can be at a rate of a cone every ten seconds or so when they are really busy. To get your soft whip into its cone takes a lot of energy. On an existing machine, the diesel engine drives a shaft which is routed through the bulkhead of the engine bay, under the passenger seat and into the base of the ice cream machine.

That spinning shaft is then connected to an air-conditioning compressor and the 'beaters' that constantly stir the ice cream mix via a series of pulleys and belts that transfer the drive to the right part of the machine and also convert it to the correct rotational speeds needed by the different parts.

It is a clever system, but the belts are not very efficient. Every belt and pulley loses drive energy through slippage and friction, increasing the energy consumption and creating heat in the cab. The condenser (or radiator) that cools the refrigerant, which cools the ice cream, is also inside the machine and all of that extra heat adds to the cabin temperature, which makes working inside the van, particularly on a hot summer's day, pretty uncomfortable.

My cunning plan was to first remove as much mechanical inefficiency as possible and then work out the simplest way to spin everything that needed to be spun with an electric motor powered from a battery pack.

I enlisted the help of my friend, John, from Vanchilla. John is a refrigeration engineer who once helped out on *Wheeler Dealers*, diagnosing and re-charging the air-conditioning system on the 1995 BMW 840Ci. You have to have special qualifications to work with refrigerant gases and, although technically I am qualified for automotive applications, you still need a lot of specialist equipment, which John has, alongside a wealth of invaluable experience.

It was obvious to me and, I believe, very important, that my electrical solution should be a kit that could be fitted to all new machines but also be retro-fitted to convert every existing ice cream van. Also, mobilers, as mobile ice cream van drivers are known, are such busy people, it needed to be a conversion that could be done with minimal down-time. Hopefully, this would mean that, instead of having to wait 10 to 20 years for the industry to convert to electric as old vans are gradually replaced by brand new ones, we could start upgrading old vans immediately.

Inside the cab of an ice cream van, the ice cream machine sits on a pedestal which is full of the shafts, belts and pulleys that drive the mechanism. So that any van could be converted without messing with that particular mobiler's custom interior, my entire kit would need to fit inside the machine, only leaving room underneath for the batteries. I needed as much space as possible inside the machine, so John stripped the plumbing down to a minimum so we could re-route the coolant pipes to fit around my new components.

Removing the inefficient belts and pulleys would help, but the biggest issue was the refrigeration system, which, with the current setup, is trying to cool down the ice cream with its own hot air. It occurred to me that we could cool things down inside considerably by running the condenser outside of the van, in the engine bay. John sourced me a more modern condenser that was both physically smaller and much more efficient. It made a huge difference.

Having done a few calculations and a lot of sitting and staring at the space available, I sourced the perfect motor and got busy with the CAD. In this case, to begin with anyway, that meant Cardboard Aided Design – a brilliant and hilarious 'concept' coined by Nick and Richard of Bad Obsession Motorsport. Sometimes, good old-fashioned bits of cardboard, pen and scissors make solving design problems even easier than on a computer.

Once I was sure we were on the right track, I digitised all of the parts and perfected the design in Solidworks and then got busy building the prototype for real. As always, it took a few late nights – all right, a lot of late nights – and a few false starts to get everything to work, especially the 'quick conversion' procedure. Once everything was in place, John re-connected all of the plumbing and I connected up the batteries. It was finally time for an 'electric' ice cream.

I turned on the machine and it sprang into life, whirring away as it started cooling and stirring. The prototype was working exactly as it should. I pulled the very first ice cream ... it was particularly delicious, the sweet taste of diesel-free success.

This summer will see some converted vans being tested in the field (or at the beach) and then the production roll-out will be in the autumn. Hydrocarbon-free ice cream for the world will be served shortly!

Years ago, I was one of a number of 'ambassadors' for a chewing gum brand. There was a cricketer, a street luger, an urban wakeboarder, a motorcycle racer and myself. I was supposed to be the cerebral counterbalance to the extreme and sporty. As part of the activity we were asked the, completely serious, question: 'What is your power animal?'

As you might expect, there were lions, tigers and dolphins, and I'm pretty sure no puppies or kittens, but all I could think of for myself was a butterfly – one flap of my wings can change the world.

Anyway, enough of this bollocks, I'd better crack on ... !

ACKNOWLEDGEMENTS

I would like to thank Imogen; my wife, my life and the woman who helped me become the man I am today. Thank you for helping me write this book, keeping my ramblings organised and my digressions within reason, and for sifting through my hoarded photographic memories.

Thank you also to the inimitable Lorna Russell for making this book happen and her editing team at Ebury for minding my Ps and Qs in every sense and cutting my words and pictures down to size to fit inside the covers of this book.

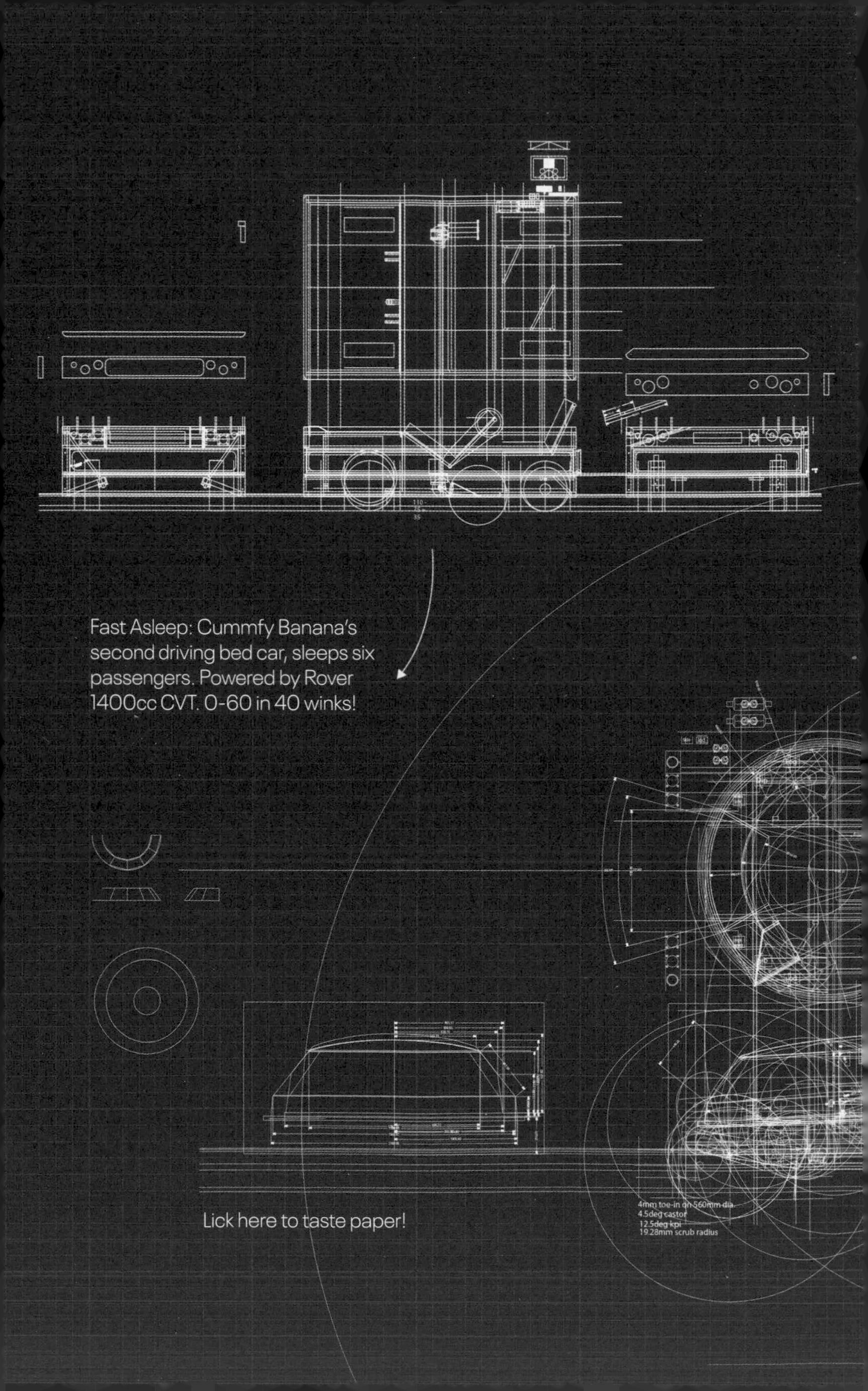

Fast Asleep: Cummfy Banana's second driving bed car, sleeps six passengers. Powered by Rover 1400cc CVT. 0-60 in 40 winks!